HOW TO GROW SCIENCE

UNIVERSE BOOKS
381 Park Avenue South
New York, N.Y. 10016

We are pleased to enclose
a complimentary review copy of

Title: HOW TO GROW SCIENCE

Author: Michael J. Moravcsik

Publication Date: December 2, 1980

Binding: cloth   Price: $ 12.50

LC#: 80-17469   ISBN#: 0-87663-344-0

We would appreciate two copies
of your review or notification
of your broadcast review schedule.

For more information contact:
Phyllis Henrici, (212) 685-7400

HOW TO GROW SCIENCE

Michael J. Moravcsik

A PUBLISHERS CREATIVE BOOK

Universe Books • New York

Published in the United States of America in 1980
by Universe Books
381 Park Avenue South, New York, N.Y. 10016

80 81 82 83 84 / 10 9 8 7 6 5 4 3 2 1

Book and jacket design by Sidney Solomon

Printed in the United States of America

Library of Congress Cataloging in Publication Data

Moravcsik, Michael J
How to grow science.

Bibliography: p.
Includes index.
1. Science—Philosophy. I. Title.
Q175.M79 501 80-17469
ISBN 0-87663-344-0

CONTENTS

ACKNOWLEDGMENTS

I*T SHOULD BE* evident to the reader that I am deeply indebted to the thousands of scientists I have met in the course of my scientific career. They and their work are the subjects of this book as well as the active or passive sources of much of its content. I consider myself fortunate to belong to such an interesting and stimulating community which extends over all countries of the world.

At the same time, I also want to mention another scholarly community from which I learned much that contributed to this book. It is the budding, varied, and colorful group of people from many traditional disciplines who work in what has come to be called the science of science, or the discipline that studies science as a human activity. They contribute, to a large extent, the perceptions and assessments of what science and scientists look like from the outside, thus complementing the internalist view that one gathers from the scientists themselves. It is my hope that popular books like this one will stimulate a broader interest in their work.

Going back further in time, I would like to record my gratitude to four people who played such a central role in helping me to become a scientist. They are my parents and two of my early teachers, Gyula Kilczer and Geza Toth, and I dedicate this book to them.

Although I had been considering writing such a book for a long time, I must acknowledge my specific indebtedness to my friend and colleague, Dr. R. S. Bhathal, the director of the Singapore Science Centre, a museum and center of amateur science activities that few other countries can match. The five articles I wrote for the *Singapore Centre Bulletin,* at the invitation of Dr. Bhathal, served as the specific stimulant needed to decide to write this book.

As the writing went along, I discussed aspects of it with a large number of people. Among them, the names of Mario Bunge, Freeman Dyson, Darrell Leamy, Edith Moravcsik, Francisco Palacio, Derek de Solla Price, and Philip Solomon stand out in particular through their cogent comments, ranging from high praise and recommendations for specific improvements all the way to the suggestion that my writing such a book should be abandoned altogether.

But a manuscript is not a book, and in connection with the transformation of the former into the latter I want to record my pleasure and satisfaction over my interaction with Sidney Solomon of Publishers Creative Services and Louis Barron of Universe Books. Their design and editing were done not only expertly but also in a spirit of friendly collaboration which made it both easy and enjoyable for me to review the manuscript with them and improve it. In this case, the proverbial battle between the author and the publisher was nowhere in evidence.

PREFACE

A GOOD BOOK is written because the author is bursting with something to say. The internal motivation, the conviction of having something worthwhile to create and share, is the best reason for doing anything, whether it be writing a book, undertaking scientific research, or climbing Mount Everest.

This book intends to be a good one, and so I should perhaps explain why, in a personal sense, I felt compelled to write it. There are at least four reasons.

First, I have been a practicing scientist for more than twenty years, and besides feeling a personal excitement about the results of science, I have been increasingly captivated by the marvels of science as a human activity. I seem to be in good company, since Johannes Kepler wrote: "The way in which man has come to understand celestial matters appears to me hardly less wonderful than the nature of the celestial events themselves." In recent years, therefore, as well as doing the usual type of research *in* science, I have also been involved in research *on* science, thus adding systematized knowledge about the workings of science to the empirical experience about it that I have automatically accumulated in the course of my career as a scientist. The picture thus gained has interested me so much that I would like to share it with many others.

Second, I have found in countless conversations with hundreds of people on five continents that the famous two-culture syndrome—that is, the separation of scientists from the rest of the world—exists not only as far as the *results* of science are concerned, but also with respect to the understanding of *how* science is done. Virtually all coverage of science in the public media (newspapers, books, magazines, radio, television, lectures) focuses on the popular representation of scientific discoveries, of facts uncovered by scientific research. This is indeed fascinating to many laymen, and should be pursued. And yet, even after many years of such reporting, laymen remain almost completely mystified about *how* such discoveries were made, *how* scientists work, *why* scientific research is undertaken, how science and technology are connected, what scientists need, and many other matters pertaining to the *how* and not the *what* in science. I feel, therefore, that a book sharply focusing on this neglected aspect of science should have an important place in the dialogue between scientists and the rest of the world. I am confident that just as the public likes to know about the lives, practices, thoughts, methods, and motivations of sports figures, movie stars, political personalities, and other people of high visibility, a discussion of scientists from this point of view will also be interesting.

Third, and somewhat related to the second reason, I strongly feel that the mystique surrounding scientific activities is, in the long run, detrimental for science and scientists. Science can survive only if at least some part of the public provides support for it. The present attitude of the public toward science is a curious mixture of adulation, fear, confidence, mistrust, admiration, and misunderstanding, and this hybrid atmosphere is not a good assurance for continued, steady support. Instead, it tends to breed alternations between ardent support and hostile withdrawal, and such a pattern makes the pursuance of scientific research very difficult. I also feel that one of the main reasons for this strange set of emotional attitudes is the lack of a rational understanding of what science is, how it is made, and what scientists are like. Thus a book that tries to unravel the mystique may perform a needed service.

Fourth, the book may also be helpful to people around the world who are involved in the management of science. Although some of these science managers are in fact scientists or former scientists, most of them are not, and their background often lacks sufficient exposure in science. The exposure is likely to be deficient not so much in knowledge about nature that science has generated—after all, it is not difficult to take a few elementary science courses on the way to becoming a science manager—but instead in the knowledge and experience about how science is made. To be sure, reading this book or any other book can by no means substitute for personal experience in science-making—which, in my opinion, should be a prerequisite for anyone who becomes, or intends to become, a science manager. Yet it is possible that a book like this will help to strengthen the basis on which science managers and administrators make their decisions.

I have written this book for anyone who is curious. There are no technical prerequisites for reading it. Even a person who never had more exposure to science than reading the daily newspapers and attending a nightmarish science course in secondary school should find it easy to go along, providing that he or she is interested in the subject.

The book consists of relatively short chapters, each of which discusses an aspect of science. The chapters can be read independently of each other, although occasionally there are references to discussions in other chapters, so that an orderly reading through the entire book is perhaps preferable.

Four appendixes discuss special topics that are connected with but not necessarily integrated into the rest of the content.

The author of such a short book can only hope to whet the reader's appetite and to direct him to turn, for detailed and possibly more involved treatment, to any of the references listed and briefly described at the end of the book, and, through them, to the large literature on science, which, in turn, is given in the bibliographies of the books listed.

This book is an experiment. Its success can be gauged only by the reaction of the readers. The mere number of copies sold is only a feeble indication of its impact. I would much prefer to receive comments, critical or otherwise, from readers so that my next experiment can be based on additional experience.

May 1980
Institute of Theoretical Science
University of Oregon
Eugene, Oregon

1
WHAT IS SCIENCE?

SCIENCE IS SOMETHING that people do. It is a human activity whose objective is learning about the world around us. That science is such an activity and not just an abstract set of laws will be emphasized and illustrated throughout this book. It is a central point both in the growing of science and in the understanding of how one can grow science.

The product of science is knowledge about the world. In fact, the word *science* comes from the Latin *scio*, meaning "I know." In present-day use the word *knowledge* has a double meaning: (1) Do you know how long the Golden Gate Bridge is? Yes, you say, it is 9,266 feet in length. Thus you are in possession of a piece of information that you know. (2) In the ancient Greek expression "Know thyself!" the word *know* does not refer to being able to recite facts; instead the expression means "Understand yourself!" In this sense *know* stands for comprehension, perception, understanding, an ability to give an explanation. When we say that knowledge is the product of science, we mean knowledge in both senses of the word; but the second sense plays a much more significant role in the development of science.

The Family of Science

There are many ways of trying to learn about the world around us, and correspondingly there are various ways in which the word *science* is used, some of them rather broad and loose. People sometimes talk about Christian Science or the occult sciences; sometimes they refer to political science or, more generally, to the social sciences. Some of this talk is perhaps motivated by a desire to share the prestige of those sciences—the natural sciences—which during the past decades and centuries have shown an astounding amount of progress and a startling list of achievements. What distinguish these various sciences from each other are the methods used to acquire new knowledge. We shall examine in Chapter 3 the method used in the natural sciences—that is, those sciences that explore aspects of the world that pertain to nonliving objects as well as to the physiological aspects of living creatures.

In contrast, the social sciences (sociology, political science, economics, history, anthropology, and others) explore the nonphysiological aspects of man, both as an individual and in a group setting. The methods they use to search for the truth are, to an overwhelming extent, quite different from those used in the natural sciences. In Christian Science or the occult sciences, on the other hand, the methods are different both from those in the natural sciences and from those in the social sciences.*

To emphasize the difference in methodology, these various sciences are sometimes also distinguished by calling some of them hard and some of them soft. In general, the sciences called hard have a less ambiguous methodology, with an ability to provide a quantitative (not only qualitative) characterization of phenomena, and are consequently more cumulative (as is explained in Chapter 3).

This book concentrates on the natural sciences, and hence

*Artists also claim to reveal the truth, and their methods differ from all the previous ones. Their claims are by no means false: To a sympathetic, experienced listener or viewer, Franz Schubert's Piano Sonata in B-flat, Op. posth., El Greco's *The Burial of Count Orgaz,* and Hermann Hesse's *Siddhartha* all provide knowledge and understanding.

from now on the word *science* will be used to mean them. Books describing the other sciences are quite different from this one, though not less interesting. Many have in fact been written, but only a few of them describe the natural sciences as a human activity.

The natural sciences are sometimes further subdivided into physical and life sciences according to whether they investigate inanimate or living objects. Thus physics, chemistry, geology, and astronomy, among others, are physical, while botany, zoology, human physiology, and bacteriology are life or biological sciences.

The subdividing goes on, moreover. The physical or biological sciences are further classified according to the particular type of phenomena under scrutiny. Astronomy deals with objects outside the Earth, chemistry studies the various ways atoms can combine into molecules, geology is concerned with the properties of the Earth, and so on.

But even this is not the end. There are subfields even within each of these scientific disciplines. In physics, for example, some people study the properties of solids and hence are called solid state physicists. Others are interested in subatomic particles, and their subfield is thus named elementary-particle physics or high-energy physics (the latter because the tools of investigation rely heavily on particle beams accelerated to very high energies). Again, others study phenomena pertaining to light (optics) or sound (acoustics).

As the overall body of scientific knowledge increases and the domain of inquiry broadens, science is pursued in terms of more and more specialty fields. At the same time, this apparently rigid compartmentalization is counteracted by investigations and fields with interdisciplinary interests. Thus people who study the physical properties of biological systems are called biophysicists, and those who study the chemical reactions in such systems are called biochemists. Similarly, there are geophysicists, geochemists, astrophysicists, and many other hybrid-sounding professions; and new specialties, bridging various old and new disciplines, are born constantly.

Besides labeling the sciences according to the particular set of phenomena they investigate, it is also helpful to make other distinctions. Two such distinctions, in particular, are used frequently, and are also frequently confused with each other.

Theory and Experiment

The first such distinction is according to a division of labor among the various components of the scientific method. This method is discussed in detail in Chapter 3, so for the moment it is sufficient to point out that science observes, studies, and describes characteristics of the systems under investigation and then attempts to order these characteristics and provide an explanation for them. The explanation, in turn, suggests and predicts the outcomes of other experiments in which other characteristics of these systems can be studied, and through such experiments the correctness of the explanation can be tested.

The division of labor takes place between two groups: *Experimentalists* design ways of measuring the properties of the systems under investigation and carry out experiments in which information is gathered about these properties. *Theorists* analyze the experimental results, attempt to create conceptual frameworks that explain these results, and then use these frameworks to make predictions for new experiments to be carried out by experimentalists. Thus we talk about theoretical and experimental science.

Let us take an example. An experimentalist is interested in exploring how an apple falls. He designs a number of experiments, using apples, stopwatches, meter sticks, photographic cameras, and other equipment, to measure the speed, the shape of the orbit, and the direction of motion of the fall of one given apple, and then perhaps he explores also whether or not a larger apple, or a yellow apple, or a wormy apple falls differently.

A theorist then takes all these experimental data and tries to order them, to ascertain the dependence of the motion on the variables, such as the size, shape, color, or worminess of the apple. Having done so, he tries to give an explanation for the regularities found in the experiment.

Let us assume that he has an idea that he thinks might

provide such an explanation. It follows from this that, other things being equal, pears would fall the same way as apples. He then asks his experimentalist friend to carry out similar experiments with pears. Lo and behold, his prediction is found to be correct: The pears behave as did the apples. Emboldened by this success, the theorist now announces his idea: All bodies in the universe experience a mutual attraction that depends only on the mass of the bodies and on the distance between them, and this dependence is hypothesized to have a certain mathematical form. If so, then the same idea that worked for the apples and pears should also work to predict how the Earth moves around the Sun. Such a prediction can be compared with the already known features of the Earth's rotation around the Sun. The agreement between prediction and observation turns out to be very good, and so the interaction between the theorist and the experimentalist proves to be fruitful and science has made some progress.

There are some scientific disciplines in which the distinction between experimentalists and theorists is very sharp, and few people are active in both experimental and theoretical work. This tends to be true in the best developed sciences such as physics, astronomy, and chemistry. In other sciences, such as in various areas of biology, the distinction is less sharp, partly because the theory is much less developed, and thus a given person may do observations and experiments, and at the same time also provide some theoretical support for his own experimental work. Yet the differentiation between theory and experiment is often useful—for example, because the working conditions and material requirements of the two groups are often quite different.

Basic or Applied?

The dichotomy between experimentalists and theorists is the first of the two distinctions mentioned earlier. The second distinction is not in terms of the function performed within the scientific method, but rather in terms of the aim, the purpose of the scientific research in question. Why we do science is discussed in greater detail in Chapter 2, so it will suffice here to mention that some of the justifications for doing science are in terms of the inherent interest and beauty of science, while others are in terms of the utility of science in other human undertakings, notably in technology. According to whether the purpose of a piece of scientific work is to satisfy curiosity or to provide a basis for utilitarian objectives, we can talk about basic (or pure) science and applied science.

It should be amply evident from the foregoing that the dichotomies of experimental versus theoretical and of basic versus applied have nothing to do with each other. One is made in terms of the methods used in the investigation, the other in terms of the purpose of the work. In particular, both experimental or theoretical sciences can be pursued either as basic research or as applied research. In our previous example, both the experimental and the theoretical work pertaining to falling objects can be motivated either by a thirst for knowledge for its own sake or by a desire to design rocket travel between New York and Calcutta.

Even though the distinction between these two dichotomies is eminently clear, it is shocking and astounding how many people confuse them and believe that theoretical science is basic science and that experimental science is applied science. Such a confusion reigns not only among people with no direct interaction with science in their formal job assignments, but even among officials and other persons who are in daily contact with science and may even take part in the management of science. Such a confusion can have some highly detrimental consequences when it comes to decisions about the input for scientific work, or to the evaluation of scientific research already performed.

Having cleared up this confusion, let us turn back to the dichotomy of basic versus applied. While the distinction between these two types of motivations for doing science is clear in terms of their definitions, in actual practice the picture is considerably blurred. There are a number of circumstances that contribute to this.

First, we must ask: Whose motivation should count when we classify a piece of research as basic or applied? Take an example, that of one of the prominent industrial research institutions in the United States, the Bell Laboratories. If we were to ask the company that finances its operations why it provides such support, the prevailing reply would be that the company hopes to remain at the leading edge of communications technology by having in-house, frontline research in those areas of science that appear to have some connection with such technology. Thus the research in the laboratory would be called applied from that point of view. At the same time, we might interview some of the physicists in this laboratory about *their* motivation for doing research. The reply then might indicate a predominantly basic motivation; that is, the researchers would claim to be engaged in this research because they find it exciting and are fascinated by the new knowledge that emerges from it. Thus it is quite possible for a scientific activity to be basic and applied at the same time, depending on whose evaluation we rely on.

The distinction between basic and applied research is hazy, however, for other reasons too. Besides the difference in motivation between various people at a given time, there is also the question of whether we want to emphasize the consequences of science in the short or the long run. X-rays were one of the most significant physical discoveries utilized in medicine. The original motivation for the investigations that led to this discovery was, however, entirely basic. The research had to do with the esoteric preoccupation of studying gas discharges in semi-evacuated glass containers, seemingly completely unrelated to diagnostics in medicine. In fact, if there had been a Council for Applied Research in Medicine in 1894, probably the last thing for which it would have provided research support would have been Wilhelm Röntgen's obscure experiments with these glass bulbs. Yet, when Röntgen discovered X-rays a year later, it took only another year or two before the new discovery was utilized in actual clinical diagnostics. What appeared to be exclusively basic research at the outset suddenly turned into most powerful applied research.

The potential for esoteric and seemingly abstract discoveries to turn eventually into economically far-reaching applications was well recognized by a number of great scientists. When, for example, Michael Faraday, the 19th-century English scientist, was asked, after one of his lectures on newly discovered phenomena of electricity, "That was very interesting, but will it ever be used for anything?" he replied, "Sir, one day you will tax it!"

X-rays do not constitute an isolated example. In fact, all scientific discoveries that are deemed significant by the criteria of basic research (that is, those that represent a considerable deepening and broadening of the our understanding of nature) have also turned out to be most significant in their applied impacts, although the latter might have become evident only some time after the initial discovery. One can therefore say with considerable confidence, based on this historical set of precedents, that any scientific work that rates high as basic research will eventually also have a major applied impact. This statement cannot be proven, and it is conceivable that exceptions will occur in the future, but on the basis of past experience such exceptions appear highly unlikely.

We see, therefore, that a second reason for the blurring between basic and applied science is that all good basic science eventually turns into important applied science. Thus again, we can have both motivations in existence side by side, one perhaps for the shorter term, the other more for the long run. This coexistence of the two motivations also holds for research that is applied in the short run: If it is of high quality by the criteria of applicability, it often also excites and fascinates the researcher and appeals to his sense of curiosity and beauty, and hence this research is also basic besides being applied.

Finally, the two motivations may be simultaneously present in the same person. We seldom do things for one single reason, and our motivations are often complex and composite. Thus a researcher may be attracted to a piece of research both by its intrinsic value and by its potential application. This heterogeneity of motivations further contributes to blurring the line between basic and applied research.

The distinction between basic and applied research is even hazier if we consider the effects of various scientific subdisciplines on each other. A research area that is predominantly basic may have an influence on another scientific research area which, in turn, may have applications. Such indirect links, if taken into account, make almost any research area simultaneously basic and applied.

Yet it would be counterproductive to erase all distinctions between these two motivations. Some balance in the overall research activities of a country, for example, should be preserved between the two types of work. This means that some, perhaps arbitrary, definition of the difference between basic and applied research should be maintained. We might say, for example, that applied research is scientific work the predominant justification of which on the part of the sponsor of the research work is its expected applicability within the next ten years in a specific area of technology or some other human activity. In those terms, then, an allocation of resources can be made for a certain amount of basic research and a different amount of applied research. How much each of these should be pursued is discussed in Chapter 8.

Technology and Science

Nowadays when one hears the word *science* in public discussions, it often occurs in conjunction with the word *technology*. Political leaders may state that science and technology are important for the country, and one may encounter councils of science and technology. In fact, there is a vague feeling in everyone's mind that science and technology are somehow related or perhaps overlapping—one justifying the other, and one the consequence of the other. Yet the distinction between the two and the relationship between them are seldom clearly delineated, and as a result, horrendous mistakes in decision-making, accompanied by false expectations and unwarranted accusations, can take place. Should pollution be blamed on science? Can we institute a ten-year crash program at the end of which a cure for cancer is guaranteed to emerge? Is the landing on the moon a tribute to science? Will added governmental support for university science departments help our foreign trade balance? The answers to these and many similar questions depend crucially on understanding the differences and relationships between science and technology.

So what is technology? We saw that science can be described well in terms of its product: Science produces knowledge about the world around us. Similarly, technology can also be described in terms of its product: Technology is an activity the product of which is a procedure for making something—a prototype of a gadget, a manufacturing process, an invention, or a patent. It is quite clear that this product is different from that of science, since science results in the rather abstract commodity of knowledge, while technology results in the more tangible products of procedure, prototype, process, or gadget.

In his book *Disturbing the Universe*, Freeman Dyson put the difference in the following perhaps overly simple but very graphic way: "A good scientist is a person with original ideas. A good engineer is a person who makes a design that works with as few original ideas as possible."

It is therefore evident that applied science and technology are very different activities. One provides knowledge, perhaps with the motivation of applicability, and the other provides the actual way of doing something new. People who are likely to be involved in applied science differ from those who are engaged in technology, since they have different skills, different mental characteristics, and different personality traits. There are, of course, exceptions, but on the whole very few scientists are also successful technologists, and vice versa.

The confusion between applied science and technology is also very common, again not only among the public at large, but even among people with a direct concern for scientific and technological matters. For example, in a recent national paper of Great Britain, written for a worldwide United Nations conference on science and technology for development, the words *applied science* and *technology* are used apparently interchangeably, thus obscuring some of the crucial issues in international cooperation in science and technology.

The distinction between science and technology can be further illuminated and, at the same time, the relationship between the two can be illustrated by a brief historical survey.

Until about one hundred fifty years ago, science and technology had practically no ties between them. That was so because up to that time technological gadgets made use mainly of macroscopic components (that is, components the sizes of which were comparable to the size of human bodies, not much smaller and not much larger), and they utilized knowledge only about the part of the world that is immediately accessible to our senses. This point requires some elaboration.

We human beings can directly perceive information about the world only through our primary senses of hearing, sight, touch, heat sensing, and smell. The type of information that can be easily received through these senses is determined by the dimensions of the human body. For example, we can easily see objects whose size is roughly comparable with the size of our body, that is, objects varying between, say, 0.01 centimeter and 1 kilometer. Similarly, we can perceive time durations that are roughly comparable to the time durations characteristic of the functioning of the human body, that is, durations between, say, 1/100 of a second and 100 years. We have direct access to a temperature range between, say, −40°C and +100°C. We can directly sense objects weighing between, say, 0.1 gram and 1,000 kilograms.

These ranges represent only an infinitesimally small part of nature. In size, we have come to know, through indirect methods to be discussed below, of objects that have a dimension of 0.000 000 000 000 1 centimeter and of objects that have a size of 1,000,000,000,000,000,000,000 kilometers. Again, by indirect means, we talk in modern science about time durations of 0.000 000 000 000 000 000 000 001 second and of 10,000,000,000 years, of temperatures of 0.000 001° above absolute zero and of 100,000,000°C. It is clear that the natural phenomena that we can directly perceive through our senses are indeed a minute part of what exists in the world.

One hundred and fifty years ago technology was based only on this minute part of natural phenomena, and furthermore, since these regularities of nature could be explored by trial-and-error procedures using our primary senses, systematic knowledge of these phenomena as embodied in science did not need to be utilized. Thus simple machines using cogwheels, levers, water power, and other everyday components could be invented by simply trying various possibilities and seeing which one worked.

The lack of interaction between science and technology at that time was also due to the fact that science was in a relatively rudimentary state. The evolution of science was also influenced by the fact that certain natural phenomena are directly available to our senses while others are not, and thus the areas of scientific disciplines that first evolved were naturally those that could be explored through direct observation with our senses. Mechanics and heat in physics were among these. Thus science at that time could hardly provide technology with any knowledge that could not also have been acquired by the would-be inventor by trial and error, and thus invention by trial and error was used in preference to science-based technology.

In the last one hundred fifty years the situation changed drastically. Having, on the whole, exhausted the most significant inventions that could be based on immediate everyday phenomena, and having acquired also the most significant items of knowledge about everyday phenomena, both science and technology proceeded into domains of phenomena that are more remote from our everyday experience and our everyday senses and that have dimensions that are significantly different (in size, time duration, temperature, and mass) from our own bodily dimensions. As soon as this switch occurred, technology was unable to rely on trial-and-error methods, since such a manipulation is virtually impossible with phenomena that we cannot directly perceive with our senses. In those domains of nature we had no intuition acquired since early childhood, an intuition that helped us crucially in the previous trial-and-error efforts. Technology therefore had to rely to an increasing extent on the systematized knowledge science could provide about these new types of phenomena. This change was already evident in technology that dealt with electricity, which began in the second half of the 19th century, and the trend escalated quickly when applications of atomic physics, nuclear physics, microbiology, geology, and other modern areas of science began to appear. Today, virtually all significant new technologies are strictly science-based. This is not to say that empirical ingenuity, and even a certain amount of trial-and-error tinkering, has no place at all in technological efforts. Scientific research generally establishes the broad outlines and the basic features of technological applications, and thereafter further development work takes place in which empirical knowledge has an important role. It is true, however, that without a science base, modern technological efforts would be doomed.

Here are two illustrations of the decline of the trial-and-error approach: If someone were provided with wooden and metal rods, nuts and bolts, welding facilities, nails, and tools, and were given the task of fabricating a reaping machine, it is reasonably likely that he would be able to do so within a reasonable amount of time. Perhaps his reaping machine would not be efficient or economical, or it might not last very long, but he would be able to approximate the original objective to a functionally satisfactory extent.

Consider now, in contrast, the same person provided with a cardboard box containing bits of copper, insulators, glass, germanium, and wood, and also with some facilities to shape and mold these materials. He would then be given the task of fabricating a transistor radio. Unless he knew something about solid-state physics and the properties of semiconductor transistors, he might try the various combinations of ingredients and the various ways of molding materials for millions of years without ever hitting, by pure trial and error, on the right combination that resulted in a transistor radio.

Today, therefore, science is the basis of technology. But the relationship also works in reverse: Today technology is the basis of science. That is so because in order to explore new phenomena in nature, increasingly remote from our everyday experience and from direct observation through our senses, we need equipment to create the phenomena that are not likely to occur by themselves in our everyday environments, as well as to measure the properties of these unusual or remote phenomena. The further removed the phenomena of interest are from our everyday experience, the more complex equipment we need to recreate the circumstances of those phenomena and to convert their signals into signals that we can directly perceive with our senses. When we measure the flow of electrons in a semiconductor, we need involved equipment to create the conditions under which electrons can appropriately flow through appropriately prepared chunks of germanium, and we also need a chain of converters to turn the flow of electrons into a signal or an instrument dial that we can see with our eyes. This instrumentation requires technology so sophisticated that much of the instrumentation used in modern science must be custom built through a close interaction of scientists and engineers. Thus yesterday's science leads to today's technology, which in turn is indispensable in creating tomorrow's science. This is a point to be pondered by those who wish for a new world in which the "bad" effects of technology are eliminated by stopping technology, but in which, at the same time, science would continue to make great strides in order to benefit man materially and spiritually. There is no such world, since modern science and technology live in a symbiosis that cannot be partially eliminated without killing both components.

In this symbiotic relationship, whenever new scientific advances are made, we can expect a corresponding progress in technology also. The two occurrences are not quite simultaneous, however. The time delay has to do partly with a lag in information transmittal, and partly with a certain time required to be inventive even in the possession of new knowledge. The first of these two delays comes about because it takes time to become aware of the technological importance of a new scientific discovery, described, for example, in a scientific journal that technologists ger erally do not read. Even within the scientific community itself, there are time delays in communication. That is even more so when knowledge must be transmitted from scientists to technologists. It is no wonder that, hoping to accelerate the transmission of new knowledge from science to technology, Bell Laboratories wants to have frontline scientific research done within its own walls.

The time delay between a scientific discovery and its technological application depends on a number of circumstances, but, on the average, the delay has been growing shorter. In the second half of the 19th century some of the major discoveries in electricity had to wait fifteen to twenty years before they were applied merely as prototype gadgets. In more recent times the delay has been more like five to eight years, as exemplified by the transistor, the nuclear reactor, and laser applications.

A word of caution may be in order, however. By technological application we mean a prototype of a gadget based on a scientific discovery. If we then want to consider the actual, industrial, large-scale manufacture of this new invention, a further delay may be expected, especially if the gadget is massive and involves a large financial investment. With the transistor, this second delay was relatively short, but with nuclear fission plants, considerably longer. Similarly, we can expect that nuclear fusion power plants will not generate a significant percentage of our energy supply before 2010 even if the scientific problems are solved in the next five years and the technological challenges of creating a prototype fusion reactor are also overcome before 1990. Similarly, even if the scientific research still needed is successfully completed in the next decade and the technological problems of manufacturing solar panels economically and on a large scale are solved, it is most unlikely that large-scale solar power plants (as distinct from home water-heating solar collectors) will provide a significant amount of energy before 2000.

Having clarified some of the basic concepts of science, technology, natural science, social science, hard science, soft science, physical science, life science, experimental science, theoretical science, basic science, and applied science, we can now return, in the next chapter, to the original statement that science is something that people do, and ask the obvious question: Why do people do science?

2
WHY SCIENCE?

A*BOUT $20–30 BILLION* is spent around the world annually on scientific research, and if auxiliary activities surrounding science are included, the sum is even larger.

In this large undertaking there are about 1–2 million people, called scientists, who, after many years of strenuous training, devote their lives to "doing science." What motivates them to devote themselves to this task?

Every year, people the world over spend countless hours reading books, magazines, newspapers, watching movies and television, and attending lectures that deal with science, even though their own livelihood, profession, and occupation do not require this. What induces them to do so?

Science—A Many-Faceted Thing

Well, why do we do anything? For some human activities, the answer is simple. The man who operates a street-cleaning machine does so because it is generally agreed that streets should be clean, and because he needs an income. For some other occupations, however, the answer is much more complex.

In such cases we may find that whether we ask about society's motivation to support it or an individual's motivation to practice it, we get, instead of a single short answer, a whole list of reasons.

Why, for example, is there a societal, collective motivation to pursue science? Ask an engineer and he may say that in his work he uses scientific results and hence he wants them to be available. A doctor may say that science makes curing people easier. A general likes science because it helps in creating weapons. To a philosopher, scientific knowledge of the world is an important input in our processes of thought. A patriot may state that science is an excellent area for countries to compete in and for his country to excel in. Others may simply say that learning about the world is exciting, and hence we should do it. Science helps lawyers by supplying forensic techniques and aids factory workers by creating the mechanization responsible for higher pay for shorter hours with leisure afterward enhanced by gadgets like television. A similarly varied list can be elicited by asking scientists themselves why they are engaged in science.

To some people there is something sinister about such a long list, especially if it contains items that appear clearly inconsistent with each other. How is it possible, ask purists, that science is favored by those who make war and also by those who cure the sick? Or by management and labor alike? Or by Conservatives and Liberals? Or by democrats and dictators? Or by Arabs and Jews? Some of these justifications for science must surely be insincere, propagandistic, Machiavellian.

Not at all. Such doubts are a result of one-dimensional thinking discussed in Appendix A. In fact, a pluralistic, multidimensional answer is a sure proof of strength, an indication that science is an extremely important activity. Humanity, after all, is very heterogeneous. That is evident if we compare various countries—or look even within a country—and find many groups of people with widely differing aspirations, aims, likes, and dislikes. Dictators sometimes would like to make it appear that their countries are united, homogeneous, and unambiguous, but none has been able to make this appearance credible, as, for example, events in the Soviet Union and the People's Republic of China have demonstrated. If, therefore, a cause, an activity, or a costly human involvement aspires to attract a broad base of support, it must have a very multidimensional, kaleidoscopic, composite set of justifications, containing what purists would regard as many inconsistencies.

If we thus look at the motivations for "doing science," we will find a perhaps bewildering variety of components. And yet, whether we consider societal, collective motivations or individual ones, we can create a certain amount of order among them by classifying them into three main headings. These are:

—Science is the basis of technology.

—Science is a human aspiration.

—Science is an influence on man's view of the world.

This chapter, therefore, discusses motivations in terms of these three categories. Although there is considerable overlap between societal and individual driving forces, it is illuminating to keep them apart. So let us begin with the motivations of society for supporting science.

The Symbiosis of Science and Technology

That science is the basis of technology needs little further mention here since their relationship was discussed in Chapter 1. Something might be added, however, about why society wants technology. There are both utilitarian and nonmaterial reasons.

In the last three hundred to four hundred years in Europe and in European-influenced areas of the world, technology has proven to be the greatest single force in making lives richer and easier. The doubling of life expectancy, the drastic cut in working hours, the enrichment of diet and shelter of an average person all result from technological developments. Perhaps even more important, technology has vastly broadened man's horizon in daily living. Long-distance communication via telephone, radio, and television have enlarged the vision of people; low-cost and rapid travel has added to the variety of personal experience almost anyone can acquire in a lifetime; reproduction of masterworks of art and music have made a once esoteric activity into virtually a mass experience; and so on. The list is endless, and impacts both on our body and soul. All this is quite evident, or at least it used to be before the blinders of fear and confusion, manifested in the various anti-science and anti-technology trends, brought about a distortion that magnifies the problems encountered in the technological evolution and ignores the benefits. The fact remains, however, that even with the problems, 20th-century living in Europe and the United States is incomparably preferable to 16th-century living in the same places, even if one compares today's general lifestyles with the style of the elite in the 16th century.

But technology is not all utilitarianism. There are many to whom technology is fascinating for its own sake. After all, technological gadgets are simply pieces of "man-made nature," and thus admiring a complex automobile engine is not so far removed from taking delight in an anthill. Yet, at least among the upper middle class, and especially in the intellectual, academic portion of it, being a bird watcher is an in-thing, but being a race car tinker is déclassé.

There is, however, also a more tangible, more objective aspect of technology, which appears to be much more independent of values and tastes. It seems that in the everlasting alternations, conflicts, clashes, and interactions of civilizations, technology is also a means of survival. It is not a guarantee for it, because many other factors, some material and some not, combine to determine which civilization, which culture, which country, or which movement survives and which goes under. On the whole, however, throughout history, civilizations with superior technology have won out over those that were more backward, more stagnant in this respect. Reference here is not only to the various arms races throughout the centuries. Arms represent only one aspect of technology. In the final analysis what matters is the overall technological capability and the richer, broader lifestyle that can result from it.

Science as an Aspiration

Let us now leave technology and turn to the second heading of societal motivations for supporting science, that described by the phrase "Science is a human aspiration."

What is an aspiration? We must have asked ourselves many

times: What is the purpose of life? This is another of those questions that may have multidimensional answers that differ from person to person. Similarly, we can ask this question with respect to a group of people, a country, or the whole of humanity. Again, at various times in history, and from different respondents, one could get widely different answers. Serving the glory of God, enhancing the might of the nation, stimulating the arts, harmonizing interpersonal relationships, becoming paramount in scientific exploration or in technological inventions are among the many possible answers.

Interestingly and significantly, the one answer not given is that the purpose of an individual is merely to eat, sleep, and procreate, or that the raison d'être of a society or of humanity is merely to take care of people's materialistic needs. Even the most confirmed hedonists have intellectual and spiritual justifications for their point of view.

This is important because it would appear that individuals and societies *must* have nonmaterial aspirations to remain creative, energetic, and satisfied. Indeed, our view of history bears this out also: We remember civilizations as well as individuals not mainly for their material aspects but for their accomplishments in nonmaterial areas. Only specialist scholars will be able to state what the material standard of living was under Alexander the Great, or under Jefferson, or in the Old Kingdom of ancient Egypt, and biographical details about the preoccupations of Newton, Beethoven, Columbus, or Shakespeare with food and shelter might, to most of us, be extremely boring.

So we all live for something beyond the purely material aspects of the world we find ourselves in. And these collective aspirations change with time. In the 20th century, scientific exploration is among the prominent aspirations of mankind. Excellence in science is *per se* (that is, even apart from the technological consequences) a major arena for international competition, and having great scientists is a matter of national pride. In Denmark, Niels Bohr, who might be called (using sloppy but picturesque terminology) the father of atomic physics, is a national hero; and Albert Einstein has become a revered symbol for all of humanity.

It is curious how scientists themselves, and many politicians, misjudge the strength of science as a human aspiration, and believe that the only way to appeal to the common man is to refer to the technological, material benefits science can bring about. Even the simplest everyday experience argues against such a narrow view. I travel much, not only on airplanes but also on buses or, outside the United States, on trains. When, in the course of a conversation, the passenger next to me finds out that I am a scientist, his first reaction usually is something like, "Oh, you must be very smart. When I went to school, I almost flunked the science classes." (That this is a misjudgment is discussed in later chapters.) A second reaction, however, closer to our present topic, is a statement like, "Well, just a little while ago, I read (or saw) something fascinating about black holes (or viruses, or elementary particles, or the movement of continents). Could you explain that to me? It must be great to be able to work on things of that sort." But I have yet to meet a person who, when learning that I am a scientist, commends me for being responsible for his color TV set.

All of us are born with a broad scope of desires, dreams, and aspirations, very few of which can be realized directly in a single lifetime. It is, however, possible to convert many of these indirectly by being part of the support of (and a spectator to) others, specially suited and trained, who attain those objectives. Watching Neal Armstrong step onto the moon, or an expedition reach the summit of Mount Everest, or Hank Aaron on a baseball field, or Jacques Cousteau in the depth of the oceans, or Sviatoslav Richter at the piano brings us excitement and pleasure, in part because our secret unfulfilled dreams appear to materialize through another human being. Similarly, the discoveries made by scientists, who are, after all, fellow human beings (even if regarded by some, erroneously, as superhuman beings), relate to everybody's aspirations.

Recently a senator visited one of the large national laboratories in the United States which is almost completely devoted to frontline research in high-energy physics. At the end of the visit, the senator turned to the director of the laboratory and said, "Well, and what is your laboratory doing for the defense of the United States?" to which the director replied, "Nothing, Senator, nothing at all—except helping to make the United States worth defending."

Science Looks at the World

The third heading for the societal motivations for supporting science has to do with science as an influence on man's view of the world.

Those of us who live in Western civilization have become so used to science as part of our outlook on the world that we tend to forget and ignore it. As a cultural and intellectual force, however, the scientific revolution, which started some three hundred years ago, has drastically changed our world view, but the only easy way to realize this today is to travel in those parts of the world not yet reached by the scientific revolution. Indeed, in many of the developing countries, science is in the lives of only a selected few who practice science or whose education has brought them in contact with the periphery of science.

That is not to say that science is a culture by itself. Science is simply a cultural element, which, although powerful, is limited to only certain areas of the rich assortment of ingredients that go into making a culture. Science is not concerned with spiritual value systems, it does not determine interpersonal relationships, and it does not strive to answer many why-type questions that philosophers, religious leaders, or politicians would ask and try to resolve. For this reason science does not replace or destroy cultures *per se*. That science could be very successfully integrated into the cultural development of Japan demonstrates that science is not a pseudonym for Western civilization.

Yet science has many assumptions, methods, and views that can make a powerful impact on any civilization with which it makes contact. Here are a few:

The first is illustrated through an account of research in Nepal on common views regarding nature, natural phenomena, and the knowledge about them. The researchers (an American physicist and a Nepalese psychologist) went from village to village, on foot, over a period of two years, asking questions such as, "If you want to find out about the explanation of rain, or earthquakes, or the eclipse of the moon, how do you go about this task?" The answer invariably was, "I look in a book or I ask an old man." The researchers then asked, "And if you cannot find it in any book, and no old man knows about it, what then?" The answer was definite and clear. "That cannot happen." In other words, knowledge was a closed domain, everything to be known was already deposited in writing or by word of mouth, and hence the road to knowledge was the diligent memorization of information in books or in the company of wise men.

The point of view taken by modern science is diametrically opposite. It claims that we know and understand only a tiny fraction of what is to be known and that therefore the road to greater knowledge is in the appropriate formulation of new questions and in the skill and imagination in devising ways of looking at the world to obtain answers to these new questions. This is, in fact, what scientists do, and they can demonstrate the success of their method by the long list of knowledge that was closed to us, say, two hundred years ago but that is no longer so.

It is not difficult to see that the difference between the two views of knowledge—the scholars would call them two theories of epistemology—is so enormous that it is bound to cause also a correspondingly drastic difference in philosophies and in practical, pragmatic views. The way change is viewed in the two frameworks is a consequence of this difference.

When knowledge is a closed domain, it is reasonable to assume also that there is no way we can improve on what happened before us, and that the world is likely to be more or less the same in the future as it was in the past. Thus any change will be greeted with suspicion, as an unnatural phenomenon. Agricultural methods used by the grandparents will be regarded as the natural method to be used also by the present generation, and values, customs, lifestyles of the traditional kind will be regarded as *the* standard to adopt.

By contrast, in a civilization where knowledge is considered an open field, and where it is accepted that history has an arrow, development a direction, and humanity a trend, change will be accepted as a natural element. There are conservatives among us who might decry a particular change (the Concorde instead of the *Queen Mary,* the computer questionnaire instead of a chat with the city clerk, jeans and long hair instead of coat-and-tie and crewcut, nuclear reactors instead of smoking coal-fed plants), but even these people expect and to a large extent accept the inevitability of change.

A second important feature of science that has made a great impact on our view of the world is its method of generating knowledge. The scientific method is discussed in greater detail in the next chapter, but for the moment it is enough to know that science claims to have a systematic way of acquiring knowledge, a method that is accepted by the overwhelming majority of people working in science, a method that can arbitrate disputes to an astounding extent, a method that can assess the degree of progress made in science and that in fact makes it possible to talk about progress.

The implication is that not only is knowledge open, but we have a way of actually enlarging our knowledge, little by little making constant progress. Furthermore, science has demonstrated that with the new knowledge and understanding, we can, to an increasing extent, influence our environment and our future. This last point is of enormous significance. Many traditional world views have strong elements of fatalism, in which human fate is totally dependent on forces outside ourselves and beyond our control, and hence, as he is tossed around on the sea of life, man is reduced to passivity and reflection.

It is no wonder, then, that the civilization first and most affected by the scientific revolution has gathered unparalleled strength and determination to change the world, and has achieved, as a result, a doubling of the human life expectancy, a manifold extension of man's horizon and sphere of action, an unprecedented equalization of the available resources between the rich and the poor of this civilization, and a novel lifestyle where struggling for mere survival has come to consume only a small fraction of the time available to the common man. Change was discovered to be possible, hence change was brought about.

Finally, the inherent optimism of science is a pivotal factor in our modern world view. In science we always assume that the problem we are working on *has* a solution. No scientist in his right mind would do research on a problem that he does not believe has an answer. When we study the confusing array of elementary-particle reactions by means of giant accelerators and huge bubble chambers, we are convinced that there is some beautiful and simple law that governs the behavior of these particles. Similarly, we engage in cancer research in the belief that there is in fact a way of interfering in the natural processes so as to avoid the development of cancer in a particular living being. This element of optimism is so prevalent that no scientist ever stops even to think about it. Imagine a scientific paper starting by affirming that the problem under scrutiny has a solution. The author would be laughed out of the community, and the editor of the journal would simply strike out the "trivial" sentence containing this affirmation.

Yet such an assumption, though commonplace in science, is not at all commonplace in other walks of life. In fatalistic societies, the assumption is almost the opposite—namely, that practically no problem has a solution that can be promoted by man. But even in contemporary Western life, many social, moral, and psychological problems do not seem to have guaranteed solutions, and perhaps indeed they do not. In this respect science cannot offer universal optimism—that *all* problems mankind faces have solutions that can be arrived at by energetic and systematic applications of the scientific method. This method may or may not be applicable to all types of problems, but it is applicable to a large domain of problems in which our optimism has been borne out by past successes without an exception. As a result of this feature of science, we can consider certain problems as settled and solved, and thus go on to other, novel, perhaps more difficult problems. In contrast, in many other areas of human endeavor we keep working on the same old problems indefinitely, providing contributions and insights, but not solutions.

In view of all these conceptual elements of science that have a great bearing on our world view, science is regarded as a "good bet" for society to support, since it is likely to create attitudes that benefit society.

Why Scientists Do Science

Now we turn to the motivations of the scientists themselves, which partly overlap with general societal motivations, but in some other respects are different.

There are many forces that drive scientists which, too, can be grouped into three categories.

The first heading can be called *introverted, internally fueled motivations*. Scientists derive esthetic satisfaction and pleasure from pursuing scientific research and from discovering laws of nature—feelings akin to those one receives from being engaged in art, either as a practitioner or as a spectator. These esthetic feelings are independent of whether or not a particular law has been known before, just as one derives pleasure from listening to or performing a musical composition, even though it has been around for one hundred fifty years and has been performed countless numbers of times.

Somewhat connected with esthetics, scientists are curious about the laws of nature, and this curiosity is satisfied by scientific exploration and discovery. This is again independent of the priority question, and is simply a personal trait the scientist was born with and the latency of which was converted to an explicit motivation through exposure to science. Why someone should have such curiosity is a meaningless question.

A third internally fueled motivation is an urge in most of us to convert talent into accomplishments. The ability to do so is an important part in our attainment of serenity. The urge is quite general, and in scientists it manifests itself in striving for understanding and discoveries in science. When we make progress on such a personal level, we are satisfied, even proud of ourselves.

These three motivations are strictly personal, and would also function if a scientist did his work in complete isolation from others, so that he would not be able to derive outside recognition or advantages from his work or be able to wonder about the priority question.

The second heading for motivations of individual scientists may be called *self-oriented within a social context,* and they pertain to the relationship of a scientist to his peers and to the community he lives in.

For example, like many other people scientists enjoy competition. The stimulation of this competitive instinct is widespread and is the basis of most games, sports, professional infrastructures, and political moves. In science this competition is particularly interesting since the participants can come from anywhere in the world, and everyone has a fairly even chance of winning, influenced only by the personal talent and drive of the participant. We will see later that this statement is a bit too idealized in real life, but at least compared to many other types of competitions in the world, it is not very far from the truth.

In the same vein, scientists strive for priority—that is, for being the first to make a discovery. The fuel for this is not only mere competition, but also the extra esthetic satisfaction of being the first in an uncharted territory.

Desire for peer recognition is another general motivation that is shared by scientists. Like everyone else, scientists like to be admired and praised for their achievements. Peer recognition, in addition, is very important for a scientist because it also validates his discovery. Since agreement between theory and experiment, the main criterion of the scientific method, may not be achieved unambiguously within a short time, a scientist derives strength and high morale, during the time when a new discovery is being tested, by knowing that fellow scientists regard his discovery as correct and important.

The previous three motivations relate a scientist to his fellow scientists. Other factors also connect him with society at large. They are social prestige, influence, and financial advantages. Again, these are universal elements which thus also manifest themselves in the case of scientists.

Finally, in the third heading, we may list the altruistic motivation of *contributing to humanity*. This contribution may be in terms of knowledge for its own sake or in terms of technology derived from knowledge. Again, most occupations share the possibility of such an altruistic justification, and science is no exception.

Determining which of these motivations is the most important is difficult since the scientific community is highly heterogeneous and various scientists would place different weights on the items in this list. Also, motivational research is notoriously difficult. If we looked only at the motivations that appear to be most conspicuous in everyday life, we would surely get an erroneous priority list, since such appearances reveal only those motivations that are challenged to the largest extent, and not those that are truly the strongest. For example, it is difficult to deprive a scientist of deriving esthetic satisfaction from his work, hence this motivation remains implicit and does not appear often in the sociological records of the scientific community. In contrast, priority is very often challenged, and appears to loom high on the horizon. In fact, some sociologists of science attempt to explain scientists almost exclusively on the basis of a craving for priority. While their conclusions are apt, they are far from being the whole truth, and in a highly pluralistic and multidimensional situation a fraction of the truth can be just as misleading as untruth.

Even though it is difficult to place relative weights on the various motivations for scientific activity, one could perhaps venture two remarks. One is that the situation is in fact multidimensional—that is, in most scientists many different motivations are present with roughly equal weights. The other observation is that I have never met a creative scientist who did not have a strong dose of the internally fueled, introspective motivations mentioned under the first heading. It seems, therefore, that curiosity, and personal joy over exploring science and achieving new insights into it, represent a necessary, though perhaps not sufficient, basis for being a productive scientist.

This is not a surprising observation, since it holds for almost any human activity. The best practitioners of any endeavor are those who are driven by forces that stem from their own makeup as human beings and are not mere reflections of the desires, values, and rewards of the outside world. Only with such an internal conviction is it possible to put in long hours, spend long years, face reverses and disappointments, and take bold risks through unconventional or unpopular moves. In the end, one's relationship with oneself counts the most, and hence it is this relationship from which strength must come.

3
HOW SCIENCE IS MADE

IN *CHAPTER* 2 we stated that science has a powerful methodology that makes progress possible, arbitrates scientific disputes, and is accepted by the great majority of scientists. What is this magic set of rules?

That there is indeed some such magic in science can already be gleaned from the way some people outside science use science to further their cause. Consider the following hypothetical, yet typical advertisement: "Scientific tests have shown that Triplicyn reduces pain three times more than the leading pain killer."

Two elements should be pointed out. First, the ad speaks of scientific *tests* and not of the *opinion* of a scientist. (Other situations when opinions of scientists are advanced as prestigious pieces of information are dealt with in Chapter 7.) The implication is that a Scientific Test (in capital letters) is a superhuman entity resulting in unlimited knowledge and an unerring, unambiguous, impartial judgment—that the way to come to such a powerful conclusion is to test (that is, to ask nature a question and receive an answer). In effect, through a scientific test, so the ad seems to imply, nature itself talks to us.

The second element worth noticing is the reference to a

quantitative result. The ad asserts that Triplicyn is three times more powerful than its competitor. To be sure, what it means to reduce pain "three times more" is left for the reader to judge; and in reality, the statement probably lacks any precise meaning. Nevertheless, the advertisers believe that adding a numerical tag to the declaration of superiority enhances its effect. In any case, people are assumed to know that science somehow deals with numbers, hence mentioning such quantitative details helps to establish the scientific authenticity of the statement.

Although couching the ad in scientific terms is in this case mostly eyewash, the implied perception of the "scientific method" does not miss the mark widely, as we shall see in this chapter.

The Four Hallmarks of Science

Science has four attributes that determine its character to a very large extent: objectivity, universality, collectivity, and cumulativeness.

As implied in the sample advertisement, science is claimed to be *objective*—that is, the results of science are thought to be independent of the particular characteristics of individual scientists, so that different individuals investigating the same domain of nature will eventually come to the same conclusions.

But is this really true, and if so, in what sense and to what extent? The statement might be disputed by all sorts of people. For example, a philosopher might ask how a person knows that the world he beholds (including all other people he interacts with) is "really" there and not only in his mind. On that level, the philosophical debate about the objectivity of the world and hence of science is perennial, and scientists, acting as philosophers and espousing their personal views, may even participate in such a debate. Science, however, does not claim to have jurisdiction over this debate, considers it outside its range of interest, and, most important, considers it irrelevant from the point of view of the scientific method. Instead, scientists concentrate on whether science (whether "real" or just an appearance, a figment of the imagination) has or has not an unambiguous, generally accepted set of results and set of procedures to attain those results.

If we look at the problem from the point of view of the existence of such a consensus within the scientific community, we discover that indeed, there is such objectivity in science. Everyone who works in science subscribes to a set of criteria that can determine whether or not a scientific statement is true; and other criteria (having to do with the status, nationality, religion, sex, race, color, age, financial means, and political ideology of scientists) are declared irrelevant in judging the truth of such a statement.

Perhaps an anecdote will drive home the extent to which status is subjugated to this objective methodology in science. When in 1957 two young physicists suggested that certain peculiar observations can be interpreted if we assume that the laws of nature are not the same for right-handed systems as for left-handed ones, Wolfgang Pauli, then one of the most influential figures in the worldwide physics community, declared: "That idea is nonsense. God would not make the world like that!" Yet only a few months later, experiments testing the idea yielded definite results showing that indeed right-handed systems do not function the same way as left-handed systems. Under the weight of experimental proof, Pauli had no choice but to admit that he was wrong.

The next objection comes from historians or sociologists who point to various instances when scientific debates were influenced by one or several of the criteria we just declared to be irrelevant. They point to influential members of the scientific community who have overwhelmed young colleagues with the authority of their views, or to proponents of new ideas in science who had difficulty in getting their views accepted.

This objection is, in some ways, well taken, but it applies only to situations in the short run. When a new set of phenomena is investigated, experiments are still in progress, a comparison of theoretical predictions with experimental findings is still tentative, the choice among alternative proposed theories is still uncertain, and the set of criteria mentioned above has not existed long enough so that a decision can be made as to who is right and who is wrong, then there is a period of murkiness during which extraneous factors can exert an influence. It seems, however, that, just as in fairy tales, in science also truth and virtue eventually triumph, and the ideas and results that are right according to objective criteria receive recognition. This difference between the short-term and long-term situations is discussed in Chapter 6.

In part as a consequence of this objectivity, science is *universal.* No matter where and by whom science is produced, the results are the same. Furthermore, the means by which the results are obtained are also the same, and this permits scientists to exchange views without any obstacles even while they are in the process of obtaining new results.

The evidence for such universality is plentiful. Many scientific conferences and meetings are international, and at such meetings two scientists, who have never met before, and with very different national, cultural, racial, religious, and political backgrounds and affiliations, can carry out, instantaneously, an animated, productive, smooth discussion about scientific matters, provided that their specialty interests in science overlap sufficiently. The only barrier may be language, but even that is not very serious, partly because communication in science does not depend crucially on great verbal skills, and partly because English is the universal language of science—so much so that when, some years ago, China and the Soviet Union were still on good terms, scientific guests from one country communicated with their hosts in English.

Partly as a result of this objectivity and universality, science is also *collective:* Scientific work involves interaction between scientists, and the results obtained by one scientist help another scientist toward further results. In addition, there are areas of science where research is done not by individuals working singly but by large groups of researchers, all of whom contribute simultaneously and in a coordinated fashion to the same piece of research. (Such teamwork, however, is not necessary for science to be regarded as collective in the sense in which we use this word.)

This collective aspect of science has very important implications. It means that communication is an indispensable ingredient in scientific work, and hence that scientists must maintain constant contact with other scientists around the world who are engaged in similar work. (This need for communication is discussed in detail in Chapter 4.)

The psychological environment in which a scientist works is another consequence of the collectivity of science. He knows that his work is tied into a huge edifice of scientific efforts, extending both in space and time, and originating with millions of other scientists, and so he can obtain gratification from making a contribution, albeit small, to this grand effort. This is in contrast with, say, a writer or an artist, whose works do not form a piece in a tightly interwoven mesh of human art, but instead must stand or fall on their own merit and on their own isolated contributions. (This and other comparisons between science and art are elaborated in Chapter 12.)

Finally, partly stemming from the fact that it is objective, universal, and collective, science is also *cumulative,* and this is perhaps its most momentous feature: The extent of scientific knowledge and understanding keeps piling up; work in science adds, at all times, to this accumulated set of results; and certain problems eventually can be declared solved, so that scientists can then go on to solving the next problem. As a result, scientific progress has an "arrow"; its direction can be unambiguously determined. In fact, the existence of this cumulation and of the arrow makes it possible to speak meaningfully about progress in science.

We regard this as the most momentous feature of science because we can claim without much exaggeration that *science is the only human activity that has this feature*. In all other walks of life, results and accomplishments do not pile up into an organized edifice, but instead remain side by side as different approaches to the same problems.

For centuries—even for millennia—philosophers have been presenting alternative approaches to the age-old problems that have been, are, and will be with us. Political systems come and go, and although proponents of a new one are, of course, often convinced that theirs is unprecedented and uniquely superior, a more detached judgment would discover precedents and would be unable to list objective (that is, generally agreed-to) criteria to judge superiority. Similarly, in art, one cannot claim that

Chagall is better, more advanced, broader, or more powerful than Rembrandt, or that Bartók's music is superior to that of Monteverdi. Different, yes; but what is missing is a way to set up a hierarchical set of criteria to judge advance, superiority, or power.

What we are saying is not meant to prove the superiority of science over art or philosophy or politics. To those of us who are intensely involved in music, it never occurs to feel disheartened because there is no way to prove that Prokofiev is a better composer than Berlioz. We are thrilled by the music of both composers and take delight in the variety music offers. Depending on our mood, we respond to one or another style of music, without missing the cumulativeness that science offers.

The difference, nevertheless, between science and other human endeavors is striking in this respect. Science is always on the move, "old" science has only an indirect importance on what happens in science today, and the research work of scientists is always in the forefront of the "new" science. It is this cumulative nature of science that enables it to serve as a basis of ever newer technological inventions.

The Interplay of Experiment and Theory

Having now enumerated some of the attributes of science, let us see how the scientific method actually works. It involves two steps, represented often by two different types of people, experimentalists and theorists. An experimentalist devises methods of getting information about a particular phenomenon in nature and then builds equipment to carry out these methods. He then performs an experiment, a measurement of certain aspects of nature under the particular circumstances defined by the equipment.

In setting up such an experiment, it is important to *isolate* the phenomenon to be studied from other interfering phenomena in which we are, for the moment, not interested. For example, in the case of the falling apple discussed in Chapter 1, a strong updraft might falsify the results because the fall would be affected by the force of the wind. We want to eliminate all such extraneous effects and observe only the one that interests us.

To judge what is extraneous, however, we must already have some understanding of the process we are exploring. For example, if we compare the speeds with which a light piece of paper and a heavy steel ball fall, we will find the latter to be greater. From this we might conclude that heavy objects fall faster than light ones. But this is an erroneous conclusion, or at least a statement that is not sufficiently precise to serve as a definite rule of natural behavior. To see this, we carry out another experiment, dropping a small and hence light steel ball (weighing the same as our previous piece of paper) and our previously used heavy steel ball. If released at the same time, they will also hit the ground at the same time, thus indicating that bodies fall equally fast regardless of their weight. This conclusion, of course, contradicts the one deduced from our experiment with the paper and the heavy ball.

How can we resolve this dilemma? Clearly, there must be some other property that we ignored and that plays a role in influencing the speed of the ball. From a comparison of the two experiments we might figure that this extra property is the size of the object, and further experiments would confirm this to be true. We might, however, have made a number of wrong starts in guessing what this additional property may be. For example, we might have guessed color, or type of material, or the time of the day, or that day's closing stock market average. Some of these we would tend to rule out as being obviously irrelevant, but to do so we must have at least a vague understanding of the nature of the free fall and the mechanisms that cause it. (In the absence of any such guide, someone may argue that on a day when the stock market average falls precipitously, the steel ball should fall faster. Can we dispute this argument on purely logical grounds?) All false connections can eventually be ruled out by carrying out more and more experiments in which more and more of the circumstances are varied, but since there are an infinite number of such circumstances that could be tested for relevance, in practice we must rely on our intuitive understanding of the process we are investigating.

It is clear from the foregoing that the other half of science, namely theory, also plays an important role, and in fact it must cooperate closely with experiment. From some experimental findings we *induce* certain generalities and explanations for the phenomena just observed, after which we *deduce* some further consequences from these explanations and generalizations, thus suggesting new experiments which, if carried out, could confirm or disprove our explanations and theoretical generalizations. Thus theory both requires experimental facts and suggests further experiments.

In the public mind scientific work is associated with logical thinking, by which people usually mean logical deduction, that is, going from a general idea to its specific consequences. As we see, this is one half of theoretical work in science. The other half, however, is not logic in this sense. Instead, it is the invention of new ideas, of general schemes, of broad models and explanations on the basis of partial, specific, fragmentary, approximate information. This inductive process is in many ways the most exciting and rewarding part of scientific work, since it is the most unpredictable element and one that requires the most imagination, the most genius, the most intuition. It is thus regarded as the most creative part of science, and is also most akin to creativity in the arts, as we will see in Chapter 12.

Not much is known "scientifically" about how creative ideas are born in scientists. There are, however, various accounts by famous scientists about their own personal experiences, and in fact almost any productive scientist will be able to recall anecdotes from his own life about how bright and novel ideas came to him. It is certainly a process that cannot be forced, and in fact the birth of a new scientific idea may occur at a time and place when a scientist is engaged in an activity apparently quite unconnected with the idea. However, personal commitment, internal excitement about science, and an intense curiosity probably are necessary prerequisites, and under those conditions the scientist's mind works on the problems subconsciously even when, seemingly, he is doing something completely different.

A striking example of this is a personal account given by the great French mathematician and physicist Jules Henri Poincaré. He had been working on a particularly challenging problem but the solution had escaped him. Then he was asked to participate in a geological field trip of several days' duration, and in the course of the preparations and logistics of the trip he had no opportunity to continue thinking about the problem. After one of the stops on the field trip, Poincaré relates, he was about to reboard the bus, and had just put his foot on the first step, when suddenly the solution to the problem struck him. Although he had no opportunity then and there to check whether or not this solution was indeed correct (the trip went on and he was chatting with the other participants), he felt absolutely certain about the solution. A week later, at leisure in his study after returning home, he checked the solution, which was correct. Accounts like this could be offered by almost any scientist who has contributed anything original to science.

A personal experience of uncanny similarity is related by the physicist Freeman Dyson in his book, *Disturbing the Universe:*

> I got onto a Greyhound bus and traveled nonstop for three days and nights as far as Chicago. . . . The roads were too bumpy for me to read, and so I sat and looked out of the window and gradually fell into a comfortable stupor. As we were droning across Nebraska on the third day, something happened. For two weeks I had not thought about physics, and now it came bursting into my consciousness like an explosion. . . . For the first time I was able to put [the pieces of the problem I had been working on] all together. For an hour or two I arranged and rearranged the pieces. Then I knew that they all fitted. I had no pencil or paper, but everything was so clear I did not need to write it down. . . . During the rest of the day as we watched the sun go down over the prairie, I was mapping out in my head the shape of the paper I would write when I got to Princeton.

Such creative thinking is very different from the logical deduction mentioned above. A noted scientist once said, "In order to invent, we must think aside"—that is, outside the accustomed groove of logical deduction from previous results.

Phenomenology

From the above description of the scientific method, it would appear that research consists of a reinforcing alternation of experimental and theoretical work, of the gathering of new information and the explanation of that information. This is indeed so. From time to time, however, experimental information accumulates but theorists are unable to find any explanations. Or, conversely, some theoretical advance is made, but its experimental verification through additional experiments is delayed, because of the difficulty of carrying out the suggested experiments. Under such circumstances, a third branch of activity, phenomenology, evolves. It tries to summarize the experimental data in an economical way, using whatever previous knowledge exists on the type of phenomena under investigation. Since such previous theory is not quite sufficient to explain everything, the remaining regularities in the data are simply described in terms of some parameters, without explaining what those parameters are supposed to mean.

Let us illustrate this with an example relating to the motion of planets around the Sun. Two significant quantities describing the motions are the distance of a planet from the Sun and the time it takes for the planet to complete one revolution around the Sun (the so-called period of the planet). These two quantities can be measured for the various planets. Let us assume that we have measured these quantities for Venus, the Earth, and Mars, and we therefore have the following table of data:

Planet	*Distance from the Sun (in millions of km.)*	*Period (in Earth-days)*
Venus	108	224
Earth	150	365
Mars	228	686

Even in the absence of any knowledge of the cause of the rotation of planets around the Sun, we can see that the greater the distance from the Sun, the longer it takes a planet to circumnavigate it, and in fact we also see from the table that an approximate doubling of the distance (from 108 for Venus to 228 for Mars) produces a considerably larger than twofold increase in the period (from 224 to 686). Let us now try to find some quantitative law that describes these variations. To start with something simple, we might guess that the periods are proportional to some power of the distances. Testing this guess, we observe that, for the Earth and Venus, the ratio of the distances is $^{150}/_{108} = 1.39$, while the ratio of the periods is $^{365}/_{224} = 1.63$. If indeed the just-assumed type of proportionality (called a power law) holds for these quantities, we can determine that power by asking by what power (or exponent) we have to raise 1.39 to get 1.63. The answer turns out to be 1.48—that is, $(1.39)^{1.48} = 1.63$. Now we take the ratios for, say, the Earth and Mars, and carry out the same operations. The result for this so-called exponent is 1.51—very close indeed to the 1.48 we got before. Just to be sure that our new rule works, we measure the distance and the period of Jupiter also: We get 778 million kilometers, and 4,329 days, from which the exponent is 1.50. Now we are really confident that our "rule" works, and indeed, the application of it to the other planets of the solar system predicts the relationships between distances and periods with a high degree of accuracy. We now have a phenomenological relationship. Nothing has been explained, however, since we do not know why such a power law should exist. But then, as a next step, a *theoretical* insight might suggest to us that a gravitational force acts between the Sun and the planets, which depends on the inverse square of the distance. From that theoretical hypothesis, we can then derive (deduce) that the relationship between periods and distances must go as the $^3/_2$ power—that is, the exponent we have been dealing with must be $^3/_2 = 1.50$. Now we have a theoretical explanation.

New theoretical insights often give a completely new way of looking at nature. For example, the motion of celestial bodies in the old days was visualized, following the Greek astronomer Ptolemy, as occurring on the surfaces of huge spheres, the center of which was the stationary Earth. This so-called geocentric view of celestial motion held for many centuries, even though in order to bring it in better agreement with observations, it had to be modified by imagining that the celestial objects move not on the surfaces of the large spheres themselves but on the surfaces of smaller spheres which in turn roll on the surfaces of the larger spheres, and in some cases even a very small sphere rolling on a small sphere rolling on the large sphere was needed to "explain" what was observed. Thus the scheme lost much of its original simplicity, which had been appealing. The predictive power of the theory was imperfect, however, in spite of the added modifications.

Dissatisfaction with both the lack of simplicity and the limitations in predictive power of the Ptolemaic system stimulated Copernicus, 1,400 years after Ptolemy, to offer a radically different concept for the motion of the planets—namely, that the planets (including the Earth) orbit around a stationary Sun. While definitely simpler than the modified Ptolemaic system, the predictive power of the Copernican idea was not significantly superior at the beginning. One might say that in predictive power there was a smooth transition from the old to the new system. Conceptually, however, the transition was extremely abrupt—so much so, in fact, that the new idea was strongly denounced as being against common sense ("We *feel* the Earth is stationary!") and against religious ideas ("Man, God's crowning creation, must be at the center of the Universe!").

Fighting new ideas was not confined to the distant past and to people with small minds. Einstein much of his life was very opposed to the ideas of quantum mechanics, which had developed first in the 1920s and dominated 20th-century physics because of its great success in explaining microscopic (atomic, nuclear, and subnuclear) phenomena. He objected to the inherently probabilistic nature of the new theory. His friend, the physicist Paul Ehrenfest, half-humorously chided him for this: "Einstein, I am ashamed of you; you are arguing against the new quantum theory just as your opponents argue about relativity theory." (To Einstein's argument "God does not throw dice!" another great physicist, Niels Bohr, replied: "It is not our business to prescribe to God how He should run the world!")

A new theoretical framework, a new conceptual way of looking at a set of phenomena, is called a new paradigm. Although scientists themselves never use this term for this purpose, sociologists and historians of science, who relatively recently focused on this well-known feature of scientific progress, use the word profusely.

The fact that paradigm changes occur periodically has been used by some people to argue that science is not cumulative after all. Old paradigms, they say, simply turn out to be wrong, and are replaced by new ones.

This view is dubious for two reasons. First, the new paradigms can usually be recognized as superior to the old ones in terms of generality, applicability, and simplicity. Thus, even in paradigms, there is an "arrow" to the progress of science. The second and more important reason is that as far as predictive power is concerned, science *is* cumulative, in that a phenomenon that is predictable in science never becomes unpredictable later, and the domain in natural phenomena for which we can give a practical prediction constantly and smoothly increases. In this sense, therefore, science is highly cumulative. Note that it is this second sense that matters when we consider science as a basis of technology, and that is why there has been a continuous and rapid growth in our technological ability even through many paradigm changes and revolutions in our conceptual view of the world. Paradigm changes are somewhat like the shedding of skin by a snake. It is the same snake, constantly growing, but it needs a new skin from time to time to hold itself together.

To recognize simplicity, however, is not always easy. A new idea may be difficult to get used to, yet it may be simple. But simplicity is not only in our minds. The great 20th-century physicist Werner Heisenberg said: "The simplicity of natural laws has an objective character, and it is not just a result of thought economy."

Crackpots

Scientific "crackpots" illustrate how not using the scientific method can lead to difficulties. Although their motivations and working patterns vary, most crackpots share the following characteristics: (1) They generally work on the grandest, most difficult scientific problems, such as the reform of the relativity theory, a new atomic theory, and the like. (2) They are always theorists; that is, they offer what they consider new ideas rather than carry out new experiments. (3) Although they claim to have formulated new theories, specific, quantitative predictions cannot be calculated from them. For this reason, among others, their articles cannot be accepted for publication in scientific journals. (4) Crackpots do not take to criticism. They prefer to claim that traditional scientists cannot appreciate their revolutionary ideas. (5) Partly because of this paranoiac attitude toward fellow workers in science, crackpots almost always work alone, and not in teams. One never sees a crackpot paper written by more than a single author. Crackpots seldom realize that even if their articles were published in scientific journals, they would continue to be ignored by other scientists, because their theories lack predictive power and cannot be tested by experiment.

There are people who come much closer to "doing science," and whose ideas are incorrect in substance but not necessarily in form. Then, of course, there are those (not all of whom are crackpots) whose ideas appear highly implausible when they are first advanced but become accepted later, perhaps as new experimental information becomes available. The idea of continental drift (the theory implying that the various continents at one time formed one large land-mass, then broke up and drifted apart) is an example of a theory that used to be in disfavor but is now generally accepted. These authors, however, follow the scientific method, present predictions that can be verified, and defend their views in scientific rather than personal terms. Clearly, there is a large variety of active people in the range between an obvious crackpot and a mainstream, productive scientist, and in some cases judgments are not easy to make and may have to be modified later. Yet most of the crackpot cases that reach a scientist's desk are very clear cut. Silence is the simplest response but perhaps not the best or most charitable one. Instead, I usually answer as follows:

> Dear Mr. Smith, I read your paper with interest. As you know, in science it does not matter who writes a paper, and whether he is an established scientist or not. The only criterion in science is the ability of a theory to make quantitative predictions for phenomena that can be measured. Since you claim that your theory is a new explanation of atoms, I would like to ask you to calculate, from your theory, the energy of the second excited state of the simplest of all atoms, the hydrogen atom. Sincerely yours, . . .

There is hardly ever a reply to such a letter, since crackpot theories cannot be used to make predictions. Yet the letter is fair, does not prejudge the theory on account of its origin, and simply requests the type of additional information that is common in scientific correspondence.

4
HOW SCIENTISTS COMMUNICATE

MANY *MEMBERS* of the public used to visualize a scientist as a man in a white coat and long white beard, who puttered with his flasks and chemicals and made discoveries in happy solitude, in an isolated setting, which, if placed on a desert island with no mail, telephone, and travel facilities, would work just as well.

Nothing is further from the truth. A scientist of today (whether in white coat or not, with or without white beard, and in chemistry or any other discipline) would be utterly unable to function creatively and productively without at least some communication with the rest of the scientific community.

The need for communication follows from the four dominant characteristics of science discussed in the previous chapter. The objectivity of science allows easy communication since the recipient of a transmission will not find it difficult to comprehend and appreciate its content. The universality of science provides an almost limitless audience, comprising the entire body of scientists the world over who specialize in the subject of the communication. Collectivity urges each scientist to receive news of all important scientific developments in his field of research, and also compels him to publicize his findings in order to make them as valuable as possible and in order to assure their incorporation into the overall body of scientific knowledge. This is the origin of the paradoxical statement that in order for a scientist to establish his "ownership" of a scientific result (that is, to establish his priority of discovery), he must promptly and widely share this result with the whole scientific community. Finally, cumulativeness gives a scientist a long-range incentive to communicate his result, since he knows that if he has done a good piece of research he is likely to have made a definite (although perhaps infinitesimally small) contribution to the overall edifice of scientific knowledge for the ages to come.

In the early days of modern science even great scientists sometimes had to be made aware of the importance of communication—often through the urging of their colleagues. Newton, for example, apparently proved to himself that under the influence of an inverse-square force, planets move in ellipses (thus giving a theoretical foundation to Kepler's phenomenological laws), but he had little interest in publishing this result until he was persuaded to do so by the astronomer Edmund Halley. Today the proof is included in every elementary textbook on physics.

This indispensable need for constant communication sets science apart from most other professions. A successful lawyer can pursue his career without constantly attending conferences, without regularly interacting professionally with many other lawyers, and certainly without paying much attention to legal developments abroad. Similarly a painter can be quite happy if he is provided with the means to live and a studio where he can pursue his art by himself. Yet it will undoubtedly please him if from time to time he can exhibit his works and see some public reactions to them, but his further art work will not greatly depend on such interaction with others.

This difference sometimes gives rise to suspicion and doubt in the minds of people who behold (or, even more important, financially support) scientists and their activities. Travel is particularly subject to such scrutiny, since it is conspicuous. As a scientist, and as someone involved in the building of science in the developing countries, I travel a great deal, and I see the incredulity in my friends' eyes when it turns out that I am off to a scientific meeting in Hawaii, or close to the Colorado ski areas, or in the foothills of the Himalayas. By the same token, however, people should, depending on their tastes, suspect a lifeguard on the beach, a projectionist in an X-rated movie theater, the chauffeur of a Rolls Royce limousine, and a worker in a perfume factory, because they constantly live with commodities which, to an ordinary citizen, appear extraordinary and expensive.

To illustrate the use of the various channels of scientific communication, let us return to our hypothetical scientist on the desert island and slowly relax his isolation. For the moment, we continue his personal quarantine so that he cannot go off the island and nobody can contact or visit him in person, but we are willing to allow him other ways of communication with the outside world. What, then, would he be doing?

Written Documentation

Perhaps, above all, he would subscribe to scientific journals. In this respect science is different from most other professions, where books place above journals. If we visit the working place of a lawyer, an economist, a historian, or a scholar of literature or of art, and if we monitor his needs in the libraries he uses, we will find a preponderance of books over journals. In social work, business, politics, and other professions, reports would loom high after books. In contrast, in the natural sciences journals play a much more important role than either books or reports. (We will mention later the "preprint" and its importance in the sciences, but such a preprint is not the same as the report which appears only in that form and is not published afterward in journals or books.) To a lesser extent, journals are also important in sociological field studies, in experimental psychology, and in other experimental areas of the social sciences which are more akin to the natural sciences. Journals also play an important part in medicine, which can be regarded as an applied natural science.

So scientists communicate their work primarily in journals. Scientific articles tend to be relatively short (perhaps as long as a chapter of this book)—because the language of science is condensed, economical, and terse, and much can be expressed in little space. The whole lifetime output of a very productive scientist may not amount to more than five hundred printed pages, an amount that is reached by almost any self-respecting historian, political scientist, literary critic, writer, or sociologist in the first year or two of his career.

Although each scientific article is relatively short, the number of scientific articles published every year is staggering: At present, some seventy thousand scientific journals around the world publish some two million articles yearly. At this rate, the amount of scientific information will soon get out of hand and will exceed anyone's digestive capacity. Fortunately, there is a miraculous and very effective compacting process operating in the sciences. At the time new scientific ideas are born, new theoretical calculations are made, new experiments are presented, there is much detail, much extraneous matter, much background that is being recorded and transmitted. For example, all circumstances and technical components of an experiment must be published so that others can judge the correctness of the experiment and, if necessary, reproduce it in their own laboratories. Similarly, calculations must be given in a complete form, with all steps recorded, so they can be judged and repeated by others. The context in which new ideas are presented is even more involved. Quite often when a new idea is born even the originator cannot clearly and simply present it. This goes back to the intuitive, inductive nature of new ideas: A scientist may visualize or "feel" the new idea, but as if through haze and fog; and although he is intuitively sure that the idea is correct and will work, it may take him or others weeks, months, or even years to convert this vision into a simple, concrete, clear, and specific statement that can be used to make quantitative predictions for the outcome of experiments.

As time goes on, compacting takes place for all types of scientific papers. After an experiment is generally accepted, its technical details become unimportant. In fact, after the domain of phenomena the experiment investigated becomes well understood, even the specific results of individual experiments may become unimportant—not untrue, merely unimportant. The experiment was important in its own time, but has done its duty and now can appear at best as one small entry in a table of data, or, even less than that, can be only implicitly present in a formula that describes our understanding of thousands of similar experiments.

Taking as an example the measurement of falling bodies used in previous chapters, it may be necessary at the beginning to describe how an experiment was designed and carried out; and different experiments, dealing with large apples, small apples, pears, steel balls, pieces of paper, done in the morning, afternoon, evening, done in France, in Peru, and elsewhere would be recorded in detail. After a while, technical details of the experiments would be omitted, and much of the data would be summarized in tables. Finally, the tables would also vanish, together with scores of other tables containing experimental results pertaining to gravitation, and only one little, simple formula would remain, thus compacting thousands of pages into half a line.

Not only experimental results but also ideas become compacted. After a certain time we are able to comprehend, simplify, and clarify an idea, and to get used to thinking in terms of it. Let us take an example. One of the most influential novel ideas of 20th-century physics is the wave-particle duality. According to this view, everything in the world is simultaneously a wave and a particle, or, perhaps less misleadingly, everything is a "wavicle"—a queer thing that sometimes has properties of waves and sometimes has properties of particles. All of atomic, solid-state, nuclear, and elementary-particle physics, and astrophysics, among others, depend on this idea.

When it was born, in the early 1920s, much needed to be written about it in scientific articles, much discussion ensued, and many of the contributions to these exchanges were hazy, tentative, and redundant. Now, less than sixty years later, the wave-particle duality is being taught to undergraduates in hundreds of colleges, although a few decades ago only the greatest minds could deal with it.

Compacting begins in science as the results of articles are summarized from time to time in review articles. Perhaps twice or three times the length of an ordinary article, and published in special review journals, a review article condenses the important content of hundreds of ordinary articles.

After a certain area of inquiry has "settled down" (that is, the experiments are verified), theoretical understanding is acquired, and in general the area appears to be ready for a logical and didactic (rather than chronological and piecemeal) presentation—books are written on it. Later on, such special monographs may be incorporated into more general books about a larger domain of science.

It is evident from this account that books in science provide primarily archival knowledge, overview, systematized presentation of well-settled facts and ideas, and that they also serve, in the form of textbooks, to educate new generations. This is in contrast with literature, the social sciences, and other areas where books serve as much for primary communication as for review and systematization.

Most scientific journals are refereed. When the author submits an article to a journal, it is sent to one or more members of the scientific community who are knowledgeable about the subject. Such a referee then submits an anonymous report on the paper, recommending its rejection or its publication with or without modifications. The author then can reply and argue with his anonymous referee or referees. The editor of the journal makes the final decision. If the article is finally rejected, the author is free to (and usually does) submit it to other journals. A large majority of all scientific articles submitted for journal publication eventually are published somewhere. This is in contrast to journals in the social sciences where only a fraction of the submitted articles survives the editors and referees. The difference can, at least in part, be ascribed to the objectivity of the natural sciences—that is, to the availability of generally accepted criteria for what is correct and interesting. As a result, an author can judge his own work before submission, and the referee will use approximately the same criteria to judge the article after submission.

Because of the refereeing procedure, and because of the logistical problems and the preparation time of scientific journals, it takes quite a few months for an article submitted to a journal to reach the potential readers. In some areas of science such a delay is acceptable, as compared to the rate at which the field develops, and in addition the group of researchers working in the specialty that is the subject of an article might be small enough so that informal communication through letters, telephone, or visits effectively and quickly complements the journals. In other large and fast-moving fields, however, the four to eight months' delay in journal publication is deemed intolerable, and so an additional network of preprint communication arises. A preprint is a duplicate copy of the manuscript that will be or has been submitted to a journal. It is not refereed before it is distributed, and this constitutes one of the points of objection to it by journal editors. This point, however, does not, in the least, worry the scientific community which is the recipient of the preprint. In one of the largest fields of physics—elementary particle physics—from two hundred to four hundred preprint copies are distributed. One disadvantage of preprint communication is that it is the author who decides who should receive his preprints. In such decisions the Nobel Prize winner (who does not read preprints at all or very little any longer) is often favored over a young, active, productive researcher in a lesser-known institution and in a distant country.

When preprints constitute the primary means of communication in a field, even journals become archival in that articles are seldom seen first in journals by those active in the field. Yet, because of the more organized, compact, and refereed format of journals, they continue to play an indispensable role. I therefore cannot comprehend or sympathize with the great fear of and opposition to preprints that journal editors tend to express.

Preprints are discarded a year or two after their appearance, since their content can then be found in journals. Journals, in turn, are very little used any more than, say, twenty years after their appearance, since by then the material contained in them has been compacted along the lines discussed earlier.

Researchers active in a given field are likely to know about the important developments in this field through journals and preprints, as well as through the personal channels of communication discussed below. Yet there may be a need to supplement such "frontline" information by more systematic, archival channels which can guarantee the presentation of all potentially relevant information. Furthermore, when someone begins his scientific career or wants to switch areas of research and work himself into a new field, he does not yet have the advantage of being embedded in the personal communication network of that particular scientific subcommunity, and hence may want to rely on something more formally structured.

How, for example, can someone in a given field retrieve information that was generated, say, in the last decade? As already mentioned, review articles constitute a good start. Another way is to scan the so-called abstract journals, in which brief summaries of past scientific articles are given, and the articles are author- and subject-indexed. Covering a broad area of science, a given abstract journal contains articles from hundreds or thousands of journals.

Another useful tool for information retrieval is the citation index. In it, article titles are listed, together with titles of later articles that refer to that article. This way, once one finds a relevant article—for example, through an abstract journal or from a review article—one can trace its sequels, the subsequent developments for which it was at least partly responsible. The science citation index also has become a pivotal tool in studying the scientific communication system itself, and the structure of science and of the scientific community, as is emphasized in Chapter 6.

Tools such as abstract journals and citation indexes have been made possible by the computer revolution of the last twenty years. Computers are also used in other ways to promote the communication of scientific information. Systems are being designed to retrieve relevant information from enormous data banks at commands that specify certain key words. For example, the computer might be fed the phrase "beetles feeding on rice plants," and it would yield a list of titles and journal locations of articles that deal with the chemistry, ecology, zoology, or any other scientific aspect of such a beetle. For the moment, such systems are still rather clumsy, expensive, and undiscriminating, but with time they may develop into a practically useful tool—especially in applied research, which tends to be more interdisciplinary and hence presents greater problems of information retrieval.

Scientists Meet and Talk

If we now return to our hermit scientist on his desert island, we find him still unhappy, despite our placing at his disposal all the journals, books, review articles, preprints, abstracts, citation indexes, and computer tapes he wants. Why is he unhappy? What elements in communication does he still miss?

There are at least three such elements. The first is access to an instant response, an instant feedback. To carry out a discussion, a scientific debate, a resolution of opposing opinions, to test a new idea through counter-arguments and expressions of critique, is extremely difficult, clumsy, and lengthy through written means of scientific communication.

Second, the written formats of communication are all, to varying extents, formal, finished, and substantive rather than tentative, speculative, preliminary, and informal. An unsuccessful experiment, for example, a calculation that did not yield interesting results or had to be aborted because of technical difficulties, generally cannot be communicated through any of the written channels (even though they constitute interesting information), since according to the conventions used by journals and other types of communication such articles would be unacceptable. Nor can one publish plans for experiments to be carried out, intentions to perform some calculations, or very speculative, unjelled ideas for future theories.

Third, written forms of communication lack the human element, the direct interaction with another human being, that can bring encouragement, expressions of interest, insights into patterns of thinking, can create acquaintance with personal behavior patterns relevant to research, and can serve as apprenticeship in the auxiliary activities of scientific research, such as organization or policy-making. Many of these features sound vague and abstract, but they are absolutely crucial for doing science.

I was reminded of this recently at a conference where a Nobel-laureate physicist was asked to give an informal, reminiscent account of his career. He said nothing about journals he had read, articles he thought highly of, books he learned from, or citation indexes that helped him greatly. His whole account was devoted to recalling the great physicists he had an opportunity to apprentice under and work with. This view is reflected by the great 20th-century physicist Werner Heisenberg, who states in the preface of one of his books that he "hopes to demonstrate that science is rooted in conversations."

Indeed, oral, personal communications between or among scientists have an immense effect on the progress of science. The most conspicuous of these is the scientific conference or meeting. Thousands of these take place every year; some are attended by thousands of scientists, while at others there are hardly more than a dozen participants. Such meetings feature longer as well as shorter talks, panel discussions, poster sessions, discussion sessions, rapporteur talks, and various other formats. It is unfortunate that many scientific meetings do not concentrate sufficiently on those formats that maximally exploit the virtues of person-to-person communication, and instead communicate in ways that can be duplicated or even improved through written channels (such as, for example, straightforward lectures with no time for discussion afterward). There is much traditionalism in the organization of scientific meetings, and more experimentation and reforms would greatly contribute to a better utilization of such meetings. In addition to what the meetings explicitly offer on their programs, attending scientists also have the opportunity for informal, person-to-person conversations.

At such conferences the main purpose is not to steal other people's ideas. My own experience is that while I am listening to a talk at a conference, I am also trying to find new and better ideas to accomplish what the speaker is trying to do, and somehow the speaker's efforts stimulate me to be creative. Many other scientists also have gone home from conferences with new ideas that came to them while they listened to speakers' struggling with a problem.

A variant of a conference is a "summer school" (sometimes

held in the winter!), at which prominent researchers in a specialty lecture on the most recent advanced developments in the field and the audience consists of other researchers active in that field. Such summer schools can also be used effectively for interdisciplinary interaction.

The proceedings of conferences and summer schools are usually published in book form. These proceedings, however, record only the most formal aspects of the event and cannot possibly recreate the many involved discussions and other person-to-person interactions that took place.

Much of the oral communication, however, takes place in even more informal settings. Departments or laboratories in scientifically active countries have several visitors a week from other research establishments who give lectures on their work and confer with local researchers on topics of common interest. Scientists from time to time spend longer periods (months or a year of sabbatical leave) at other institutions, in order to get acquainted with different points of view and to absorb new ideas as well as propagate their own.

But even in their home institutions, scientists spend considerable time discussing problems with each other, getting critiques on some of their new suggestions, conjectures, or proposals. In fact, a certain critical mass of scientists in one geographical location and with strongly overlapping research interests is crucial in generating creative, productive research activity. Generally, three to four people are sufficient to form such a research cell.

The telephone is also a very important tool of scientific research, as illustrated in the following example. Suppose I need a piece of information or a hint for locating a piece of information. I know that I could go down to the library, and, at the cost of a half an hour of search, probably find the answer. I also know, however, that my colleague at, say, Brookhaven National Laboratory is in daily contact with that piece of information and hence has it at his fingertips. Instead of wasting my time in the library, therefore, I pick up the telephone, and in two minutes, at little cost, I have the information.

But the telephone serves scientists in other respects also. Preliminary results of crucial experiments, important happenings at conferences one did not attend, or current developments in the efforts to provide adequate support for scientific activities can often be best ascertained through the telephone.

Another element in scientific communication—secrecy—tickles the public fancy. Since science depends on communication, and the productive capacity of a scientist and his ability to establish priority in a discovery also depend on it, there is relatively little secrecy in scientific research. Sometimes researchers do not publicize their findings until the work is completed and the results are certain, but this is due as much to prudence as to a fear of being "scooped" by other researchers.

In some aspects of applied scientific research, military or commercial considerations dictate keeping the research secret. But even that does not represent a permanent secret but only the gaining of some time over others. Once the time has come for a scientific discovery to be made, generally several scientists or groups of researchers will make it more or less independently and simultaneously. Especially since science has become a big-time activity (starting with the close of World War II), such multiple discoveries are more the rule than the exception. It should be noted, however, that gaining time on others in military and commercial situations is of great essence, hence secrecy in those areas is certainly understandable and logical.

Excessive secrecy, however, is harmful even to the scientific community involved in the particular research, since it implies bureaucratization, and because if each scientist has to justify his "need to know" for every piece of scientific information, the usual cross-fertilization of ideas discussed earlier cannot take place. It is impossible to give a blanket rule of how much secrecy is in fact necessary in a particular situation, and the judgment must therefore be made in each case by someone well acquainted with that area of science.

This chapter has discussed only communication *among* scientists. The equally crucial question of how scientists communicate with nonscientists is taken up in Chapter 12.

5
HOW TO MEASURE SCIENCE

A *STRANGE TITLE, PERHAPS*. In previous chapters we learned that quantitative measurements of natural phenomena are indeed very important in science, but why would anyone want to measure science itself—and how?

Many people, however, may find it useful to quantify scientific work. For example, educational institutions that train future scientists may want to know how big their science departments should be or are expected to be. Governments or private companies that supply funds for the maintenance of scientific laboratories may want to know how large an expenditure they must count on. Young people going into science may want to know the size and status of the job market now and for the years to come.

Perhaps even more interesting and important, those who provide funds for pursuing science may want to know how much they got out of these funds. Or an employer of scientists may want to measure the relative merits of various candidates for a job. Or a historian of science may want to ascertain the rate at which science has grown over the years. A man involved in international science policy who wants to compare scientific work in, say, Malaysia and Venezuela may need quantitative measures for both countries.

The situations described in the previous paragraph are particularly interesting and important because evaluation and assessment of scientific work are both difficult and crucial—because knowledge, the product of science, is hard to measure; yet it is also crucial that it be measured so that we can distinguish between success and failure, between excellence and incompetence, between a fruitful investment and a colossal waste of resources. Precisely because the fruits of science are not instantly and clearly evident to laymen or even to supporters of science, we must have some generally agreed-on means to measure science.

Gauging the Input

Following the method often used in economics, we may consider science as an input-output system. Certain resources are fed into the scientific infrastructure, and certain output results are achieved. The measurement, therefore, must deal with the input, and then with the output, and should also enable us to compare the two in some way to determine something about efficiency.

The measurement of input is relatively much easier. What does science need in order to prosper? Money, is the easy answer. Sure enough, money can buy many things science needs: buildings, research equipment, people's salaries, tools of communication, for example. Yet money cannot buy everything that must be fed into science to get results. No matter how much money is fed into an educational system, there is no guarantee that it will produce an Einstein. But even on a more mundane level, an expensive laboratory does not necessarily assure the type of human atmosphere conducive to scientific efforts. Neither does money guarantee a flexible, knowledgeable, sympathetic, efficient bureaucratic system that supports science. Some of these elements are discussed in greater detail in Chapter 10. For the present, the important point is that there are both tangible and intangible input items, and even of the tangible ones, only some can be guaranteed by a proper expenditure of money.

In any case, however, the tangible input factors can be measured to some extent. We can count the number of laboratories or other research institutions, list the research equipment, give a number for the size of the scientific manpower, describe quantitatively the libraries and the expenditures on scientific communications. In fact, such quantification of the more evident input ingredients of science is often seen in the reports of agencies, in national development plans, and in the annual reports of companies. Although they serve a purpose and provide some information, by themselves they are woefully inadequate even to characterize the tangible part, let alone the intangible part, of the input.

Perhaps the greatest shortcoming of this simple-minded quantification is its neglect of quality. Two scientists are seldom identical in their potential production of important scientific results. Two pieces of equipment, costing the same, can differ drastically in the care applied to their manufacture, in their durability, and in their ease of handling. Two libraries, with similar budgets, may differ from each other substantially, depending on the competence of the library staff. The effectiveness of repair facilities for scientific equipment cannot be described merely by the amount of resources fed into these facilities, because the skill, dedication, motivations, and competence of the technicians play a crucial role but are not characterized by the size of the input. The budgets, number of students, or size of the staff of scientific educational institutions tell us very little about the quality of education offered in the institution. We can give such examples for all the items needed for scientific work.

In summary, therefore, we may conclude that while we can quantify some aspects of the input of science, such a measurement can deal reliably only with the more tangible part of such an input, and even there it is bound to ignore the element of quality which can play a decisive role in the value of the input.

The problems of measuring input are, nevertheless, minor

compared to the problems we face when we want to measure scientific output.

Activity, Productivity, and Progress

To begin with, let us distinguish among three concepts: scientific activity, scientific productivity, and scientific progress. To illustrate this, let us imagine that we visit a scientific institution to ascertain how good it is.

As we tour the institute, we see experimenters puttering with their equipment, theorists busily writing formulas on pieces of paper, others scurrying around computers. If we are laymen, this is really all we can see in a scientific institute, and we shall come away with the conviction that the institute must be a very successful scientific center.

But are we justified in coming to this conclusion? A bit of reflection will convince us that this is not so. All we have seen is scientific *activity,* but we have no evidence that anything was actually *produced* that could be regarded as an output of science.

We therefore return for a second visit, this time wiser and hence in the company of a scientist who will help us to assess more than merely the scientific activity in the institute.

We do that by contacting the experimentalists and asking them about the results of their experiments. They will show us tables of data, books full of graphs, and copies of scientific papers they have published in journals. The theorists, similarly, will show us calculations, theories, comparisons with data, and, again, publications in scientific journals or invitations to give lectures on their theories at scientific meetings. Those who work with computers will show the results of their computations and the summaries of their work in reports or other publications.

To authenticate this evidence, however, we need our scientist friend with us, because only he can tell whether the experimental results make sense or are based on bogus experiments carried out incompetently, whether or not the journals where the results of the research were published are "respectable," or whether or not the reports of the work are worth anything.

Let us assume that, in all these respects, the assessment of the institute is very favorable, and so we leave, now with a more certain feeling that the institute is indeed a very successful scientific center. After all, it shows high scientific *productivity*.

But then again doubts begin to arise. What is the purpose of science, after all? Is it to produce tables of measurements, fancy theoretical schemes, endless print-outs of computer calculations, and lists of journal publications, or is it to solve problems, to enlarge our understanding of nature, to acquire knowledge with which we can modify the world around us through technology? Clearly, it is the latter, and although the former is said to further the latter, how can we be so sure that that is the case in this institute, or, if so, to what extent, in this case, scientific productivity actually contributed to scientific *progress*?

It is very difficult to answer this question, even at the cost of a third visit to the institution. Often scientific progress can be judged only by hindsight, when we already understand a certain domain of natural phenomena and hence can judge how a given piece of research carried out on the way to this understanding contributed to the final goal.

The relationship of activity, productivity, and progress can be illustrated by an analogy. Suppose we want to get from a place in a dense mountainous forest to the entrance of a gold mine. We know that there is a gold mine somewhere in these mountains, but we do not know its location. Activity then would correspond to the amount of thrashing around, the amount of hacking at tree limbs, the amount of clearing of the underbrush.

Productivity would correspond to the number of feet of trail that resulted from all this activity. Finally, progress would be measured by the extent to which the distance between our original location and the entrance of the gold mine has decreased. This latter quantity cannot be figured out with absolute certainty until we actually find the mine, even though we may have encouraging indications on the way by, say, measuring the amount of gold dust we can pan out of the creeks we cross.

So when we say that we want to measure the output of

science, we should specify whether we want to measure activity or productivity or progress. In practice, what we can measure contemporaneously, to a significant extent, is only activity, mixed with some productivity and perhaps a small amount of progress. With hindsight—that is, looking at scientific activities that took place decades or centuries ago—we may be more successful in judging productivity or progress as well.

How do we carry out this measurement? In the preceding chapter we saw how scientific communication is central in scientific work, and how the activities of a scientist are useless unless he communicates his results to the scientific or technological community. It is therefore logical to use the communication process as the gauge with which scientific output is measured.

There are basically two different ways of performing such a measurement. One appears to be more objective, more impersonal, more quantitative, and utilizes the written modes of scientific communication. The other seems to be more subjective, more person-to-person, and less quantitative, and is based on peer judgment—that is, on the evaluation of scientists through the opinion of other scientists. Since, however, the written modes of communication also originate with scientists, the two modes are not as different as they may appear at first sight.

Objective Measures

Let us first talk about the objective methods of measuring scientific output. We saw that the most important written vehicle of communication among scientists is the journal article, and so these methods are linked with such articles. Through the computer-handled abstracting journals and citation indexes, it is easy to gather statistics on the number of authors, the number of articles by a given author, the number of citations an author receives, and all these numbers can be used as measures for scientific output.

Let us ask, for example, which of two countries—Finland and Mexico, say—is more advanced in science? One way of answering this question is to count the number of scientists in each country who published one or more scientific papers in, say, 1975. This number of scientific authors turns out to be 1,075 for Finland and 479 for Mexico, from which we conclude that Finland is the more advanced in science.

Let us, however, take a different situation: In a given university science department, there are two contenders for one tenure position. The measure we used just above would not work very well in this case. If both contenders have published papers in 1975, they would be ranked equal by our measure. A measure that can have some meaning in one application may be quite useless in another.

But even as applied to countries, the measure we used is rather imperfect. It could happen, for example, that in country A there are forty highly productive researchers, each of whom writes five papers every year, while in country B no one does research after publishing his Ph.D. thesis, but the university system produces sixty Ph.D. students, each with a rather poor thesis publication to his name. Our measure would rank country B higher than country A, simply because there are more authors with one poor paper each in country B than there are authors in country A with many excellent papers each.

So let us not count only authors but also scientific papers. In this way the productive scientist will be recognized more than the colleague who writes only one paper.

This measure will also work better for the contenders for the tenure position we mentioned. If one of them is very productive and has many scientific papers to his name, he will show up better than the other who published only a few papers.

This measure of the number of publications is in fact widely used in university circles, and has given rise to the slogan "Publish or perish." In many situations it works quite well, and it clearly provides more information than the number of scientific authors. Yet it also has some grave deficiencies.

For example, Wolfgang Pauli, one of the greatest figures in 20th-century physics, published about eighty journal articles in physics proper throughout his entire lifetime, fewer than I have already published, even though I am considerably younger than Pauli was when he died. Using our measure of the number of publications, therefore, I should be placed above Pauli, which is an absurd conclusion.

The problem, of course, is that quality is not considered in a simple count of the number of articles. In recent years this might even have influenced publication patterns in physics, in that a scientist might think (and, in part, justly so) that his standing with the dean might be improved if he published many short articles rather than a few long ones, or several fragmentary articles on a given piece of research instead of waiting for the research to conclude and then writing a concise, comprehensive article.

How can we remedy this defect? Somehow we must weigh the articles by their quality. Remember that science is collective, and hence a high-quality, "important" paper represents a more substantial, more influential contribution to the overall edifice of scientific knowledge. If so, it is likely that its content will be utilized by a larger number of scientists in subsequent years. Each such utilization will leave a trace through the utilizer's referring to (that is, citing) the article he used. Thus we might think that counting the number of citations of a given article will represent an appropriate measure of importance or quality of that article.

Thus we might assess the quality of a scientist not by the number of articles he wrote, but rather by the number of citations his articles have received. Counting the number of citations is easy if we have access to the citation index.

Such citation ratings of individuals, groups, institutions, or even countries have in fact been used in various situations where evaluation of scientific work was called for. Again, there is no doubt that such citation ratings give more detailed information than do either the number of authors or the number of articles. Yet we are still far from having a perfect measure, as the following examples will demonstrate.

Suppose someone works in an area of research in which a large number of scientists are interested, and suppose this researcher published a competent but not very significant paper. Others in the field who publish papers after him will then feel obliged to cite his paper, not because it contributed substantially to subsequent papers, but just because it exists and so courtesy suggests its existence be recognized. Such perfunctory citations are quite large in number: Some studies seem to indicate that between a half and two-thirds of the citations in scientific articles are in fact perfunctory and hence indicate little about the quality of an author of a cited article or the impact of that article.

As another example, consider an article with an important new conceptual idea. If this idea catches on, very shortly it will be so much a part of public knowledge within the scientific community that there will no longer be any need to cite it by giving the location of the original article. Thus the number of citations of this article may be low, even though its impact on subsequent science is substantial.

Let me again offer a personal example. My best contribution to physics so far (by the judgment of the scientific community as a whole)—the contribution through which I am best known in the physics community—was a conceptual one, which soon became generally adopted and hence ceased to be cited. In contrast, my most-often-cited article is a piece of work I was first reluctant to publish at all—it was completed in two to three hours and was a calculation any student could have performed. It turned out to be a useful though trivial tool in the work of many other researchers, and hence it continues to be cited. Since it is not an idea but only a tool, it keeps being cited by a reference to the journal location and hence keeps contributing to my citation rating, even though I consider it a trivial accomplishment, compared to my other article, which by now hardly contributes to my citation rating.

Some problems of counting citations could be avoided if the contexts of citations were also investigated and citations were classified, for example, according to whether or not they are perfunctory. Such classifications have been explored, to some extent, and some interesting results have emerged from such work, but since the classification must be performed by hand and not by computer, it cannot be done on a sufficiently large scale to be useful in most practical applications.

By now, the reader may get the impression that any measure using journal articles is bound to be hopelessly inadequate and full of deficiencies. This is by no means the case, however.

These measures are in fact very useful for various purposes, and interesting applications have been developed with their help, both in the practical administration of science and in the study of the structure of science. Some of them are mentioned in the next chapter. It is important, however, always to keep in mind the limitations of these measures, so we do not use them under inappropriate circumstances. Furthermore, we should try not to use them exclusively but always in conjunction with other information, particularly information obtained from subjective measures, those based on peer judgment.

Peer Assessment

Suppose a university science department wants to promote a young faculty member. Besides investigating his publication list, citation rating, and other objective measures, the department may solicit the personal opinion of, say, five generally respected scientists who work in the same area of research as the candidate. It does not matter at all where these referees are located. In fact, since science is so universal, they are likely to be in five different parts of the world. Each of these referees then writes a letter of reference, sometimes rather extensive, sometimes only a page long, in which he assesses the merit of the candidate, based on the latter's published articles, appearances at scientific conferences, or personal interaction with the referee. By asking a number of referees, the department hopes that any distortions due to rivalries, personal animosities, and other irrelevant effects can be eliminated.

A similar kind of peer judgment is solicited when a department, institution, or laboratory is evaluated for its production and quality. Usually a visiting committee comes to the institution to be assessed and, for a week or so, interviews researchers, inspects their working environment, looks at their publications, and then presents a report of its conclusions, including recommendations for future improvements.

As we saw in Chapter 4, scientific articles submitted for publication to scientific journals are also assessed by such peer judgment—namely, the refereeing system.

An advantage of the peer judgment system is that it can be based not only on those elements that are evident in written communications but also on other relevant factors. For example, in an evaluation of the performance of a given scientist, his contribution to the work of his colleagues in terms of advice, suggestions, comments, or critique is of considerable importance. There are a number of highly respected members of the scientific community who themselves publish relatively little, but who are very instrumental in originating new ideas or, through informal discussions with colleagues, in promoting the effectiveness of theories or experiments. Although such interaction is quite often recorded in the acknowledgments sections of journal articles, the measures based on articles, which we discussed earlier, ignore such acknowledgments since they are not citations in the formal sense. Peer judgment is also a more effective method of evaluating the potential of a researcher in terms of his ideas for further research, in terms of his research still in progress, or in terms of his ability to follow the development of a fast growing field. These all are important aspects of a scientist's productivity, but they cannot be assessed directly from the objective measures discussed above.

Another advantage of peer judgment can be illustrated when an evaluation is to be made of an institution in a developing country. In such a country the conditions under which science can be pursued tend to be different (invariably more difficult) than in countries where science has been well established for decades or centuries. Research may progress at a slower rate because equipment, repair facilities, chemicals, and other material aids to research may not be readily available. Scientists may have to perform more of the preliminary work themselves because communication channels with the rest of the scientific community may not be ample, and hence they may not be aware of relevant work already performed elsewhere. Bureaucratic obstacles for such scientists may be even more severe than for their colleagues in scientifically well-established countries, thus robbing them of time and energy that otherwise could be spent on research. Purely objective measures cannot easily account for this difference, but an appropriately knowledgeable visiting committee can do so.

We see, therefore, that peer judgment is a complementary method of assessment that should be used in conjunction with objective methods, even though the latter appear to be superior because they implicitly involve a large number of fellow scientists while peer judgment is invariably based on a relatively small number of opinions.

Putting It All Together

Objective and subjective measures are not quite as different as they may appear on the surface, because the number of publications of a scientist and the number of citations he receives are connected, indirectly, with peer judgments. Journal articles, before publication, are judged by referees—a form of peer judgment. Citations definitely result from peer judgment, since a citation is awarded only if the citing author knows the cited article and deems it worthy (for one reason or another) of recognition. Thus objective and subjective measures should not be expected to yield very different results if applied to the same subject. In fact they do not, in many cases, although there are exceptions. Even if the conclusions of the two methods are similar, however, they do give different types of information and hence, together, provide a broader picture.

It is sometimes claimed that these measures of the output of science are applicable mainly to basic research and are misleading or unfair when used for applied research. In the large majority of cases this is not so. Results of applied science are communicated just as much as those of basic research, and in fact many thousands of scientific journals specialize in publishing the results of applied research. Similarly, peer evaluation is just as feasible for applied research as for basic research. To be sure, some military or commercial research is kept temporarily secret and therefore is communicated not through journal articles but through classified reports, in which case the knowledgeable peer group may be rather small. Still, our measures have been used, apparently successfully, in the case of such classified reports and employing small peer groups. It should also be realized that secret research is generally declassified and made public fairly soon, since monopoly over a piece of knowledge can be preserved only for a relatively short time. Thereafter, the usual application of measures can take place unhindered.

One might also suggest that a good measure of the output of applied science would be the impact such research made on technology—and in some cases this might indeed be so. In many other instances, however, such a measure would be unreliable, partly because there is a time lag between applied scientific research and its technological application, and partly because a low score based on measuring the technological output, or the output in terms of agricultural or industrial production, may not be due to the quality of the underlying science but to the inability of the country (for managerial or other reasons) to utilize applied research in technology and production, even if such research is of high quality and great potential significance.

From the foregoing discussion, it should be clear that the measurement of input is very much easier and, at least to a rudimentary extent, can be done by people outside the sciences, while the measurement of scientific output requires, at almost every stage, specialists familiar with those active in the sciences. For this reason, planners often confuse measures of input and output with each other or simply substitute the former for the latter. Thus one may read regarding a national development plan that the goal for science and technology is to produce 3,500 more scientists in the next five years. The tacit though undoubtedly unintended implication is that, in the eyes of the planners, the purpose of science is to produce scientists and not to acquire new knowledge. In fact, there is hardly any mention in any development plan of the targets of scientific output, but only of the input.

A discussion, however brief, of measuring science would not be complete without mentioning the crucial role in science of exceptional people whose impact and importance may not be adequately described by any of the measures listed above. If, for example, we want to assess the impact of Einstein on science, measures like number of articles, number of citations, and contemporary peer judgments would all be inadequate, even though they would rate Einstein very high. I said "*contemporary*" peer judgment because the main difficulty with measuring science and at the same time the most important application of it is with respect to contemporary or near-contemporary events and personalities. There is little difficulty in recognizing Newton's superb contribution to science, since this is old history and our view of his impact benefits greatly from the hindsight of two hundred and fifty years. Very few figures in the history of science are as exceptional as Newton or Einstein, but there are many more exceptional people on a somewhat more modest level, and the measures we listed, if they are applied contemporaneously, may tend to undervalue even them. As a result, scientists who are even just suspected of being truly exceptional should be supported properly so we do not stand guilty, in the judgment of posterity, of failing to recognize great opportunities for advancing scientific knowledge. It was the impact of such exceptional people that Ralph Waldo Emerson had in mind when he said, "Every revolution was first a thought in one man's mind." Perhaps nowhere is this more true than when applied to scientific revolutions.

6
THE SCIENCE OF SCIENCE

T*HAT SCIENCE,* which has become a conspicuous and massive undertaking, is subject to scrutiny, study, investigation, and inquiry is due in part to science itself. In a scientific age, the rational, "scientific" investigation of everything is in vogue. Such scrutiny, however, is also motivated by practical considerations, since any activity that consumes large amounts of material and financial resources and has a great influence on the world must be able to explain itself to the society that supports and is influenced by it.

In some respects the study of science as a human activity is not new. More than three hundred and fifty years ago, Johannes Kepler (1571–1630), the famous astronomer, said, "The way in which man has come to understand celestial matters appears to me hardly less wonderful than the nature of the celestial events themselves" (although that he did any more thinking on this subject is not evident from his writings). The history of science, however, is a rather venerable discipline, even if, for some time, historians of science contented themselves mainly with telling or analyzing the life stories of famous scientists. The philosophy of science also has been flourishing for some time, since philosophers have been captivated by the structure of the scientific method, and also because man's view of nature, as it developed in the last three centuries, has had a considerable impact on philosophical thinking.

The Birth of a New Discipline

More recently science has become the object of curiosity from other directions as well. As the scientific community grew (and it did so precipitously, as we shall see), sociologists became interested in its structure, in the interactions among its members, in how it is molded by the opinions and personalities of its leaders. And with the great rise of psychology in the 20th century, some of its investigating techniques and results were applied to the scientist, to scientific creativity, to scientific interaction.

The foregoing groups represent the academic side of the study of science. Other people, however, have been interested in understanding the workings of science for practical reasons. In an age when private or governmental bodies are asked to award large sums to scientists for research activities, those in charge of making such decisions feel uneasy about having to base their decisions only on guess and on the personal experiences of previous science administrators. In true scientific spirit, these decision-makers would like to have more objective knowledge and understanding about science as a human activity in order to improve the efficiency and impact of their operations.

Economists also are interested in science, partly for academic and partly for practical reasons. Science and technology have an immense effect on economics, hence they must be taken into account in economic theories and practices. As an example, various national development plans include a substantial section on science and technology, and since economists tend to write these plans, they ought to learn something about the structure and the features of science.

Finally, in the course of their research activities, scientists themselves constantly encounter problems and decisions that require, in addition to a technical understanding of their particular fields of science, some understanding of science as a human activity. This is particularly true for senior scientists who have been working in the field for a decade or two and who are therefore increasingly enlisted to serve as scientific leaders or as organizers and managers of science. In addition, as people grow older, they tend to think more about the whys and hows of things than younger people who often just charge ahead in their youthful enthusiasm and singlemindedness.

Thus in recent years, various kinds of people have been converging on the study of science—so much so that their combined interests constitute a new discipline, a new field, sometimes called the science of science.

In some ways the name is unfortunate, because the science of science is not a science, at least in the sense in which the word *science* is used in this book. Its researchers, its methodology, its internal structure are, at least for the moment, much closer to the social sciences than to the natural sciences. Quite possibly this will remain so, and it may be that the scientific method is actually unsuitable for the study of science itself. Only time will tell.

It is fascinating to observe how such a new discipline evolved. It began with people in various established fields becoming interested in one or another aspect of science. These interests could be pursued only if the interested person had strong credentials and hence a secure position in his traditional discipline. For example, somebody who wanted to study the sociology of the scientific community had to have to his credit some good work in other, more acceptable areas of sociology, and had to have a position (usually at a university) that allowed him to branch out into this new area. Similarly, a scientist who wanted to do research in the science of science had to be already well recognized for his contributions to his specialty in science itself, and had to have an appropriate job that allowed him to spend time on such research. In fact, often this extra research had to be done after hours, in addition to conventional research activities.

As the new discipline jelled and more and more people became interested, its existence became visible through more tangible signs. By now there is an international association for researchers in the science of science, called Society for the

Social Studies of Science (the 4S society), and there are journals (for example, *Social Studies of Science, Research Policy,* and *Scientometrics*) that devote themselves entirely to the science of science.

Universities and governmental agencies, however, have not yet formally recognized the new discipline. No university yet, to the best of my knowledge, has a department and gives degrees in the science of science; neither is there a Division of the Science of Science in any funding agency. In some ways, such conservatism is prudent. Universities should not bend with every breeze of academic and nonacademic fashion that happens to come and go, and initially experimentation and flexibility should be encouraged within universities through the individual activities of faculty members and students. Formal restructuring of departments and curriculums should take place only after a new discipline has shown its vitality, durability, and productivity for a decade or two. In the meantime, the new discipline will depend on the type of "moonlighting" described above, and on the devotion and enthusiasm of those who are initially involved in it.

A few institutions, however, have some formal entity (even if not a regular department) that deals with the science of science, or with some part of it. In the United States, for example, Cornell University has a Program on Science, Technology and Society, and Harvard has a Program for Science and International Affairs. In England the University of Sussex houses a Science Policy Research Unit; in Germany there is an Abteilung für Wissenschaftsforschung (Division for the Study of Science) at the University of Ulm; and in Canada the University of Montreal has an institute for the history and the social policy of science. Twenty years from now, many major universities may well have given formal recognition to the science of science.

Since the people now active in the science of science have come from many different backgrounds, there is no unified, central point of view in the discipline, and the various groups pursue many types of research. For example, one major division is between people who want to describe science by quantitative measures and parameters (the scientometrologists), and those who want to use more traditional research tools of history and the social sciences in general, and thus work through case studies, in-depth qualitative investigations, and by setting up of theoretical hypotheses and then trying to muster qualitative evidence for their validation. Although there have been heated and even acrimonious confrontations between the proponents of these two points of view, in practice these confrontations have been indecisive, so both approaches continue to be pursued. This is most appropriate in a discipline so new and so rudimentary. No point of view can claim superiority on the basis of past successes, hence pluralism must prevail.

The rest of this chapter provides additional examples of problems and areas where the science of science can give or has given us insight and understanding.

The "Realities" of the Scientific Method

The first of these examples returns to the scientific method as outlined in Chapter 3 and modifies that discussion to include some of the realities of human interaction. The analysis and understanding of these realities is the function of the science of science.

In Chapter 3 the scientific method was sketched in terms of the interaction of experimental findings and theoretical (or phenomenological) interpretations, explanations, and predictions. An experimentalist prepares an experiment that asks nature a specific question, and nature answers the question through the result of the experiment. Then a theorist interprets this result through a theory that can account for the results of all past experiments as well as of the present one, and in addition makes a prediction for one or several new experiments to be performed. These new experiments are then carried out, and their results either confirm the existing theory (or, perhaps more appropriately, fail to disprove it) or conflict with it. In the latter case, a new theory must be formulated to explain this experimental result.

All this sounds very objective, aseptic, mechanical, unambiguous, and impersonal and appears to imply that the method operates quite independently of the characteristics of the people who operate it.

To a considerable extent this is indeed true. Compared to the ambiguities, vagaries, subjective and personal elements, and ambivalent methodologies of other human activities, science does possess a great deal of objectivity and impersonality, especially if we follow its development over a longer period of time. Nevertheless, in order to understand the detailed operation of science, we must also study those human factors that become mixed into this ideal, objective scientific method. To illustrate what these may be, let us follow one cycle of the scientific method as described above and insert into it the human elements as they occur.

To begin with, an experimentalist must be employed and must have financial support to carry out his experiment. To do so, he generally must possess some credibility in terms of past scientific accomplishments that are recognized by others in the scientific community. Such recognition, however, can be influenced by where the experimentalist is located, whether or not he has good access to communication channels, whether he is an extroverted or introverted person, whether his personality is attractive or repulsive, and many other human factors.

Furthermore, in order to attract financial support for his experiment, he must also compete with other experimental proposals for the same funds. Making a decision about which proposal to fund depends on a factor not mentioned in our original discussion of the scientific method—the ability to guess, before an experiment is carried out, whether its results will be interesting enough to merit funding. Making such a guess is a rather vague operation in which there is plenty of room for irrelevant human factors. A similar problem arises if the experiment in question is to be performed on a large multipurpose scientific facility like a particle accelerator or a satellite. In such cases a second selection must be made among experiments already funded, since space and time on such facilities are always more limited than the number of experimental proposals.

Even the initial choice of experiment is open to vagaries. Since there are always more possibilities open than an experimentalist has time and money for, choices must be made. Some experiments would probably generate solid, certainly useful, yet not earth-shaking results, while others would be much more speculative, resulting perhaps in nothing useful, but perhaps in revolutionary new observations. Again, *a priori*, there is no sure way of making these choices, and so the personality of the researcher, the circumstances of his employment, his degree of security and self-confidence are all likely to affect his decision to some extent. In this, as well as in the other human elements that appear in this discussion, intuition, inductive reasoning, and vision of the unknown play a very large part. (This point is also taken up in the discussion of science and art in Chapter 12.)

The other protagonist in the scientific method, the theorist, also must have credibility to be able to have a firm position and funds to do research. But he also encounters other problems that are specific to a theorist. As mentioned in Chapter 3, the creation of new scientific theories involves a great deal of inductive thinking, imagination, and speculation, and is greatly influenced by interaction between theorists in informal discussion sessions or more formal get-togethers. Once a theory is formulated, it may encounter additional difficulties. It might, for instance, make predictions that are beyond contemporary experimental techniques to verify, or it might be so complicated in detail that making a precise quantitative prediction from it is mathematically too demanding an undertaking.

Finally, even if a theorist can explain previous experiments in terms of a theory that can further predict experiments that could be performed, he must "sell" this theory to an experimentalist so the latter will decide to perform the desired new experiment. Since an experimentalist always has more choices for new experiments than time and money with which to perform them, there is an opportunity here for irrelevant human factors to enter when a theorist tries to convince an experimentalist that his proposed experiment should get priority.

We can add to all this the element of competition, when two scientists work on the same problem and try to beat each other in solving it first. Such a race can be won by a scientist with more influence, a more commanding personality, a better place of work, a greater ability to attract able collaborators, or a stronger physique to put in long hours of work.

It should be very evident by now that psychology, sociology, economics, history, public administration, and other disciplines have indeed a rich new domain for research in the study of science as it occurs in practice. To be sure, many of the details listed above matter "only" from a personal point of view and probably do not greatly influence the long-term development of science, thus leaving science with the overall objectivity mentioned earlier. Other details may make some difference in the rate at which science develops but not in the direction of development and in the actual knowledge and understanding that are reached. Yet in practical science management, these details can count much, and they also contribute to our understanding of how the new scientific knowledge is created.

Scientometrics

Let us now discuss, as further examples for the preoccupations and results of the science of science, some relationships uncovered by that part of the science of science (called scientometrics) which tries to apply quantitative measures to characterize science. Chapter 5 dealt with such measures, so here we shall only summarize some of the applications and results.

For example, scientometrics has documented the extremely rapid growth of science. Such studies have shown that ever since the middle of the 18th century, there has been a surprisingly uniform expansion of science that obeys a so-called exponential law, that is, a pattern of *doubling* every fifteen years. This law holds for both the number of scientists and the number of scientific papers published. A few consequences of this law will bring home the tremendous impact of this pattern. There are now about ten thousand times more scientific papers in existence than there were two hundred years ago, and three-quarters of the scientific literature available to use today was written since World War II. Perhaps even more strikingly, almost 90% of all scientists who ever lived are alive today. Thus, in one sense we can say that the history of science consists mainly of the study of contemporary or near-contemporary science, since most of the developments have taken place recently. In another vein, the growth of the number of scientists is much faster than the growth of human population, hence the former must level off sooner or later, before it demands that every man, woman, and child on the surface of the earth become a scientist. (More about such ultimate limits of science in Chapter 8.)

This rapid growth of scientific information also affects the scientific communication network. It is indeed fortunate, in this light, that the kind of compacting of scientific knowledge discussed in Chapter 4 takes place, otherwise scientists would soon find themselves spending the first eighty years of their lives just catching up on science done previously, and new research activity would become impossible.

A second illustration of scientometric results is Lotka's Law, which pertains to the distribution of productivity in a scientific community. We said earlier in several connections that there are considerable differences among scientists as far as productivity is concerned—for example, as measured by the number of publications. Lotka's Law turns this observation into a quantitative rule by stating that the number of scientists, each with a certain number of papers to his credit in his entire lifetime, is inversely proportional to the square of that number of papers. In plainer English, this means that if in a scientific community there are ten thousand scientists each of whom published one paper in his entire lifetime, then there will be one hundred scientists in the same community each with a lifetime production of ten papers and only one scientist with one hundred papers to his credit.

We see, therefore, that high scientific productivity is very rare within the community of scientists. For every very productive scientist, it seems that we must have a huge number of others whose individual contribution is comparatively quite negligible.

A consequence of Lotka's Law is that half of the scientific literature produced by a scientific community containing *n* people is produced by a small group of people whose size is the square root of n. For example, in a community of forty thousand scientists (approximately the number of physicists in the United States today), two hundred are responsible for half of all the articles produced.

Lotka's Law is an empirical observation, deduced simply by collecting statistics on scientific productivity. The actual statistical numbers are very well approximated by this inverse square law, but the agreement is not perfect. At the low end, the law misses the number of people with only one paper each, and of course it predicts nonsense for the number of people with zero papers each. At the high end, the law predicts some scientists with an arbitrarily large number of papers (if only one considers a large enough scientific community), while in practice nobody could produce, say, ten thousand papers in his lifetime. But these are irrelevant blemishes. In the range of practical interest, say between two papers and two hundred papers each, the law works very well. It might be mentioned that the verification of this law is so far based almost exclusively on statistics gathered in countries that are scientifically advanced and where science has been established for some time. Whether it holds in the developing countries is not known. One of the interesting fields of the science of science is the attempt to explain the validity of Lotka's Law in terms of the physiological, sociological, psychological, and other factors that influence science and the scientific community.

Another area that has been subject to scientometric studies is the relationship between a scientist's age and his productivity. An adequately precise account of the situation would be too lengthy for the space available here, because the results depend on the scientific field, the definition of productivity, the geographical location of the researcher, and various other factors. If we use simply the number of articles published per year, and if we somehow apportion scientific papers with several authors, and if we average over many scientific fields, we arrive at the conclusion that productive scientists rapidly increase their productivity shortly after the beginning of their career, reach a plateau, and maintain this plateau for many years. A decline in productivity may then occur in dropping down to lower plateaus in succession, or even abruptly over three to four years.

This picture appears to conflict with the belief that in the sciences youth prevails and the most significant and greatest discoveries are made early in the lives of scientists. Actually, there is no conflict between the two results. The simple publication count does not indicate the importance of an article, and so even though the publication plateau may extend over a thirty-year period for a given scientist, it is also true that the conceptually most novel, most imaginative, most revolutionary (that is, paradigm-changing) ideas are likely to come to a scientist in his twenties or early thirties. This effect is relevant mostly to theorists, who are the generators of paradigm changes.

As a final illustration of results of scientometrics, let us explore the relationship between the total amount of science pursued in a country and the economic size of a country.

Specifically, let us consider the numbers of scientific authors in various countries who have published at all in a given year, and see if this parameter has any relationship to the gross national products (GNP) of the countries. (The GNP is a measure of the total value of the goods and services produced in a country in a given year.)

Both parameters vary enormously from country to country. The ratio of the largest to the smallest GNP in the world is about 10,000 to 1, and the ratio of the largest number of scientific authors to the smallest such number is about 100,000 to 1. But despite this huge variation in each of these parameters, we find that there is a quite well defined relationship between the two quantities, in that the two are roughly proportional to each other. (In more precise mathematical terms, the number of authors is proportional to the $4/3$ power of the GNP.)

It is not difficult to understand, qualitatively, why this should be so. As we will see in Chapters 8 and 11, the cost of scientific research per scientist and per year, averaged over all scientific disciplines, is determined mainly by the requirements of science itself and is therefore fairly universal. The cost may change from country to country by a factor of 3 or perhaps even 10, but not by a factor of 100,000. Then we can assume that (with only a handful of rather unimportant exceptions) the fraction of the GNP spent on scientific research does not vary tremendously from country to country. As we will see in Chapter 8, this statement is also true to within a factor of 10, and furthermore a country that spends a smaller percentage of its GNP on science will likely be one where the cost of science per scientist is also lower. With these two assumptions, the relationship between the GNP and the number of scientific authors follows.

Considering the huge variations in each parameter, the relationship is surprisingly good. It is far from exact, however. More detailed studies have shown, for example, that the relationship is somewhat different for scientifically well developed countries than for the developing countries. Also, especially at the lower end of these parameters (that is, when it comes to the smaller developing countries), some countries deviate from what the relationship predicts, by a factor of 3 or even 10 in either direction. Thus the GNP is by no means the only determinant of the amount of research that can be done in a country, and devoted manpower, intelligent management, good communication channels, and many other factors (or their absence) can play a great role in placing a country considerably above or dismally below the figure predicted by the relationship. More will be said about this in Chapter 13.

So much for a taste of the science of science, a new but fast growing field of inquiry, highly interdisciplinary in terms of the traditional modes of investigation, and very pluralistic in its approaches and results. A quick overview has prepared us to turn now to the practical questions of the organization, administration, and management of science, which is the subject of the next chapter.

7
SCIENCE POLICY

HEARING THE PHRASE *science policy*, a man on the street might imagine a room in Washington (or Bonn, Tokyo, Moscow, Paris, or any other capital) with a half a dozen wise government officials preparing a blueprint of what science should be done in the country over the next ten years.

There are two elements in this picture to point out in particular. One is the idea of centralization—the idea that a few people in a remote room at a given instant can determine what hundreds of thousands of scientists will do for years to come. The second is that the wise men are engaged in *planning*—that is, preparing instructions for the future. The implication is that science will then take place in the country in accordance with such instructions.

Both of these ideas, although not altogether specious, coincide only with a very small part of what science policy actually is.

Planning, Deciding, and Implementing

First, let us state that a policy is a set of actions designed to accomplish a certain objective. Note that policy is action, and not merely plans drawn on paper. In fact, we can say that any policy has three aspects: planning, decision-making, and implementation. For example, in an expedition that aims to scale Mount Everest, plans are made, with the use of photographs and maps, about which route to take; then a series of decisions is taken about the acquisition of provisions, porters, gear, and transportation; and finally the expedition puts in a great deal of effort actually to get to the top of the mountain.

Before looking at these three steps in the case of science policy we should note that much of the time the phrase *science policy,* as it is used in governmental discussions, in economic circles, among bureaucrats, and in other nonscientific environments, means science and technology policy, and in fact, much of the discussion is on technology and not on science. This unfortunately sloppy terminology creates much confusion, because many features of technology policy are quite different from those of policy pertaining to science. Inasmuch as technology is science based, there is, of course, a certain branch of each of the two policies that deals with the connection between the two. Although this is an important aspect of both policies, it is a rather small part of the whole picture. Since this book is about science, we shall not deal with technology policy except to the extent that it is pertinent to science. The bibliography, however, includes some items that deal almost exclusively with technology policy.

So on with the three steps in science policy. Can one plan for science? In previous chapters we saw that science aims at discovering new laws of nature, and it is rather unpredictable, since neither the law to be discovered nor the road leading to it is known ahead of time. Thus, it is not possible to prepare a timetable for progress in a scientific discipline as it might be in a technological undertaking such as the landing of a man on the Moon, which could be exactly budgeted, predicted, and carried through, following a rigid timetable.

Indeed, the output of science cannot be predicted. Yet, some planning for the input of science, along very broad lines, can be useful. A scientific project needs manpower, facilities, and money, among other things, and, if undertaken, it should have the assurance of these items for some years so that the project can be completed.

In an earlier chapter we mentioned the "accidental" discovery of X-rays. Much can be done to improve the circumstances for such accidental discoveries to come about. For one thing, scientists must be deeply and personally committed to and involved in research for such discoveries to materialize. Only he who seeks will find. Second, researchers must be conditioned, trained, and oriented toward recognizing the importance of unusual events. As the old saying goes, "Those who have eyes, let them see." The history of science abounds in episodes of researchers' missing an important new phenomenon even when they observe it. They may attribute the observed results to malfunctioning instruments, or to statistical error, or to their inability to isolate completely the phenomenon they wish to observe. When, for example, it was discovered in 1958 that laws of nature are not the same for right-handed and left-handed systems (technically called the nonconservation of parity), it was realized that some observations had been made at least a decade before which, instead of being ascribed to inaccuracies in observation, could have been interpreted also as evidence for parity nonconservation. The experiments in 1958 were carried out and interpreted correctly because by that time there was some theoretical expectation that parity may not be conserved and hence the experimentalists were conditioned to look for such effects. But even if there is no such theoretical guidance, a good researcher must have an awareness for spotting completely unexpected features in his experimental results.

Finally, the input for doing successful scientific research must be present if accidental discoveries are to come about. Especially nowadays, when scientific research has become quite costly, even the devoted and eager scientist, capable of spotting unexpected results, will be crippled unless he has at his disposal some means with which such research can be carried out. Thus some arrangements for providing input for scientists are essential, and in this respect planning can play a useful role.

The need for some planning, however, does not imply the need for centralization. In the United States, for example, there is no national science plan (and there never has been one) that would allocate a fixed amount of resources among the various organizations that pursue science. For one thing, only about half to two-thirds of the money spent for scientific research comes from the federal government. The rest comes from private industry, state governments, foundations, universities, and many other organizations which are not coordinated among themselves or with the federal government. Furthermore, even within the federal government, a number of major and many minor agencies spend money on scientific research, and each prepares its own budget more or less independently of the others. In each of these individual agency budgets, the total money available for research is roughly fixed, but its distribution depends on a number of factors that are not amenable to long-range planning. (More about how to give out money for science in Chapter 11.)

Similarly, there is no planning in the United States for a fixed scientific manpower quota for, say, ten years from now. Even if such a quota were desired, it could not be attained without regimenting educational institutions, public or private employers of scientists, or the location and type of scientific work—in other words, without taking measures that would destroy the very salutary scientific mobility within the country and degrade the quality of the manpower. As a result, fluctuations in manpower inevitably occur—shortages of scientists at one time, surpluses at other times.

In some other countries, central planning is nevertheless practiced, but its effectiveness is yet to be proven. In spite of this, the planning of science policy (especially in countries where hardly any science exists) takes an inordinate amount of time and effort and becomes a formalistic obsession that can easily divert attention from the much more important phases of science policy, decision-making and implementation.

Although planning could possibly be centralized, decision-making *must* be decentralized since it occurs at all levels of scientific activity. A senior scientist, choosing a certain approach to a certain scientific problem, makes a science policy decision, which will affect him and his research group. Decision by researchers in a given problem area to hold an international conference on the subject is a policy decision with possibly very important consequences.

Finally, implementation is an even more decentralized process, depending on all participants in the scientific enterprise. We must consider implementation the most important phase of science policy, since one can imagine scientific work carried out without planning and with only a minimal amount of decision-making, but without implementation nothing will happen. Yet this phase receives, relatively speaking, the least amount of attention in many countries that pride themselves with having a science policy. (More about this in Chapter 13 and in Appendix C.)

So much for the definition of the three aspects of science policy. Various aspects of decision-making and implementation form the subject of Chapters 9 through 12, thus underlining again their importance as compared to planning.

Policy, People, and Problems

Each of the three stages of science policy can be either policy *for* science or policy *with* science. Policy *for* science is policy aimed at making possible the pursuit of science. Policy *with* science is public policy in which there are scientific elements—that is, which uses science, among other ingredients. For example, policy for energy for the future is a policy *with* science, since the determination of our energy needs and of the means with which we shall fill them is a multidimensional problem (see Appendix A), with science and technology representing one aspect, one axis.

It might appear that policy *for* science would depend a great deal on policy *with* science since the public policy issues for which science is used would seem to determine the internal policy for science also. This, however, is not so in practice. If anything, the reverse is true: The internal development of science determines the type of public issues in which it will play a role. Clearly, nuclear weapons would not be a public issue if nuclear physics had not developed for reasons of its own. True enough, the *rate* of progress in science can be influenced by public issues (as we will see later in this chapter), but most of the internal science policy for science is very little affected by such issues. Thus it is possible, to a large extent, to discuss policy for science independently of policy with science.

Depending on the level and the organization, the organizers, administrators, and managers of science come from various backgrounds. The successful ones, however, are, or have been, scientists themselves. And because, justly or unjustly, scientists tend to divide people into "us scientists" and "them bureaucrats," and regard the latter, at best, as inevitable nuisances, a manager of science should appear to the scientists with whom he interacts as one of them.

But being a scientist or a former scientist is not quite enough for success. Perhaps the next most important attribute of a successful science manager is to have had a scientific career to look back on with which the manager himself is satisfied. The important element is the personal, inner satisfaction with past scientific achievements. This serenity gives the manager a basis for interaction with scientists without feeling inferior, envious, or discouraged.

As mentioned earlier, the interaction between science and technology is an important aspect of science policy. Although planning in detail for such interaction is not easy, one can create certain environments in which it is more likely to take place. Universities generally are not suitable for this, because most of them house relatively little actual technological work (as distinct from some technology education). Governmental or private research laboratories can do it much more easily, because with hundreds, even thousands, of scientists and technologists in the same institution, it is just a matter of internal communication practices to achieve some interaction between the two groups. In addition, having academic scientists serve as part-time consultants to industrial or governmental laboratories helps to bring about such interaction.

As an example of science policy decisions that are taken by individual scientists or groups of scientists, let us consider how to choose a scientific problem to work on. A bad way is to pick a problem merely because it has not yet been solved. Many problems have never been solved simply because they are conceptually uninteresting and will not bring about a new understanding of nature. For example, innumerable lines in atomic or molecular spectra have never been measured, yet we are not interested in measuring them, because even if we did so, the data would not be utilized in further scientific research or technological applications and the results would arouse no interest among other scientists. Noted for his caustic wit and blunt manners, Wolfgang Pauli, the famous 20th-century physicist, was approached one day by a young physicist with a new theory. After enthusiastically explaining his ideas at the blackboard, the young man turned to Pauli expectantly and asked, "Well, Professor Pauli, what do you think?" "Well," replied Pauli, "this is not even wrong!"

So, instead of simply selecting a problem because it has not yet been solved, a researcher should ask himself three questions:

(a) Am I genuinely intrigued by, interested in, and enthusiastic about the problem?

(b) Does the solution represent a positive step within the broader context of science? Would any other scientist in the world be interested in the solution, and would he build further research on it? Does the solution contribute noticeably to a better understanding of the laws of nature?

(c) Does the solution represent a positive step in applying the results toward an approach to a material or social problem? Will any technologist be interested in the result and be stimulated by it to develop, build, or design?

The problem is worth researching if the answer to (a) is yes and the answer to either (b) or (c) is yes. Although these criteria appear to be obvious and self-evident, their strict application could prevent much sterile research in science. We might call this an example of grass-roots science policy.

Policy and Politics

Finally, let us turn to some comments about an often

emotionally charged subject, the relationship between science policy and politics.

Of the languages I know something about, English is the only one in which *policy* and *politics* are two separate words. The German *Politik*, the French *politique*, the Spanish *politica*, the Italian *politica*, the Russian политика, and the Hungarian *politika* all mean both politics and policy. This is highly unfortunate, since it obstructs a clear discussion of the difference and the relationship between these two concepts and contributes to the predilection to mix politics into science.

But the discussions surrounding science and politics are confused for other reasons also. One-dimensional thinking (Appendix A) tends to obscure the true structure of the relationship, sloppy terminology creates artificial conflicts, and wishful thinking (often arising from ideology) spurs sweeping, unsubstantiated statements. Although it is impossible to do full justice to the subject in a short space, let us try to sketch some of its interesting aspects.

In a most general sense, we can say that politics is the process that determines who gets what, when, and how, or, in somewhat different words, that politics is the process of attaining goals and the means by which this attainment is achieved.

Taking this definition at face value, and applying it literally yet generally, we quickly come to the conclusion that virtually everything is politics. Following a similar largess of thought, we can also state that everything is science (since everything involves phenomena which are subjects of science, and all technological products are based on natural laws which science studies), or that everything is economics (since all phenomena in the world have direct or indirect economic implications), or that everything is nutrition (because the human beings involved in anything depend on food to survive). This kind of broad talk is not only another example of one-dimensional thinking but is also completely useless, in a practical sense, in analyzing problems. We must be much more discriminating in explaining what kind of politics we want to consider and what the nature is of the relationship we want to explore. Nevertheless, such a chaotic, careless statement is often heard, especially from ideologically motivated people, who then "deduce" from it something that does not follow from it at all, namely that science must be subjugated to political ideology.

So let us be a bit more skillful and subtle. First, we should remember that politics enters only when a conflict exists. If a group of people do not differ in their ideas and practices concerning who gets what, when, and how, no politics will exist.

Let us then ask the following question: Is there politics within the scientific community during the practice of the scientific method?

We have seen that while, in the long run and on a large scale, scientists have a generally accepted methodology that can objectively arbitrate disputes (and hence will not give rise to politics), in the short run and in smaller details, many human factors enter into the interaction of scientists. Hence politics can arise in such situations—that is, there can be a conflict during the process of deciding who gets what, when, and how. For example, when a decision must be made about whose experiment will fly in the satellite that will first reach Venus, scientific criteria will mix with subjective elements in reaching a conclusion. Thus, we can say, there is some "scientists' politics" in science, although in the long run it is not likely to have much of an effect on the direction in which science develops or on the results it acquires.

Now let us turn to other kinds of politics—international, national, and ideological politics—which are the activities that the everyday meaning of the word *politics* usually refers to.

- Do they influence the *direction* of development in science, the *results* that science attains, and the *methodology* science uses? In other words, are there several kinds of science?
- Do they influence the *rate* at which science develops and the *relative balance* between the various scientific disciplines at any given time?

The answer to the first question is no. We have seen that science has a strong element of objectivity and a generally agreed-on methodology which are internally determined and are immune to outside influences. An excellent proof of this is the fact that the direction, the results, and the methodology of science in various countries, cultures, political systems, religions, and ideologies are remarkably similar. Thus science in the United States, in France, in the Soviet Union, in Japan, in India, in the People's Republic of China, in the Republic of China, or in Nigeria is from these points of view virtually indistinguishable. That is why scientists who meet at conferences can interact instantaneously and easily regardless of their political and other backgrounds. Indeed, in this respect, there is only one scientific method.

Although authors of numerous historical studies have claimed that certain scientific developments were consequences of the political, social, or cultural climate in which the discoverers were embedded, I have never been convinced that these factors affected the direction, results, and methodology of science. Since historically there *was* only one science, such studies are merely speculative, and their authors have simply tried, *ex post facto*, to correlate certain factors.

The answer to the second question is definitely yes, but the analysis of this question is complicated, since some of the influencing factors are positive and some are negative.

That the answer is in the affirmative is clear from our discussion of the motivations for pursuing science. Some of the societal motivations listed there obviously fit under the umbrella of international or national politics. It is also clear that such motivations can affect the relative prominence of a given scientific discipline compared to others by enhancing or suppressing its rate of development. For example, with the rapid development of underwater travel and exploration, international politics has paid increasingly more attention to the three-quarters of the surface of the earth that is covered by water. Hence oceanography as a branch of science has received a boost from the funding agencies of almost all major countries.

On the other hand, there is a limit to how much one branch of science can be advanced without advancing the others, and this limit is mainly determined by factors internal to science. Both the experimental techniques and the theoretical, conceptual elements of the various scientific disciplines are strongly interrelated, and so a gross neglect of a certain scientific field will also have a negative effect on the progress of others. For example, the nuclear magnetic resonance method for various chemical analytical investigations constituted an extremely new powerful tool, but it could not have been developed without progress in nuclear physics.

Scientific methodology is another area where politics, and particularly ideological politics, can affect the rate at which science develops. Although science has a generally agreed-upon, internally determined method of proceeding, and this method is greatly responsible for its past successes, from time to time ideologues, who believe that their ideology must dominate and govern every human activity, try to force science and scientists to adopt, instead, a methodology dictated by ideology. Three recent examples stand out. In Nazi Germany, anything Jewish was outlawed, and the German scientific community was "cleansed" of Jewish scientists. In the Soviet Union under Stalin (and perhaps, to some extent, even today) scientists were forced to refer to the works of the "great scientist" Stalin, and scientific ideas, contributions, and results that were not "materialistic" enough for the taste of the commissars were blacklisted. In the People's Republic of China under Mao (and perhaps, to some extent, even today) scientists were forced to use, as methodology, the works of the "great scientist" Mao, as laid down in *The Little Red Book;* and to bring scientists closer to the "people," they were also forced, part of the time, to perform menial labor in rural areas.

Hitler, Stalin, and Mao are dead and their madness has been exposed, but the damage they did to their countries' science remains a deterrent against the repetition of such mistakes. German science during the 1930s and early 1940s deteriorated and its productivity plunged to a low. Soviet science, considering the available potential talent and the amount of material resources it consumes, it still underproducing, especially in areas such as genetics (which was plagued by Lysenko and his supporters). Although the Soviet Union supports a larger body of scientific manpower than the United States, in 1973 only 8% of the world's scientific authors were from the Soviet Union, while 42% were from the United States. Chinese science is in serious disarray, some scientists have been humiliated and discouraged, a certain age group of younger scientists is completely missing because advanced education was shut down during the Cultural Revolution, and scientists are still hampered by the inability to travel freely abroad to interact with other scientists. In all three instances, material resources placed at the disposal of science were very considerable—in fact more generous than in many other countries—which again demonstrates that money, by itself, cannot buy good science.

Why are ideological dictators so eager to interfere with scientific methodology? The answer may, to a large extent, lie in the objectivity of science, in scientific methodology that advocates, as the criterion for truth, comparison with observations rather than obedience to an authoritative set of commands. Such objectivity is a threat to a dictator. At the same time, in the 20th century, it is generally recognized that science, being a basis of technology, is an indirect key to political and military power, and hence the dictator knows that he must have science. This schizophrenic attitude (on the one hand wanting to possess science and on the other hand being afraid of and feeling threatened by science) can be particularly well demonstrated in the ups and downs and the 180-degree twists and turns of science policy in the People's Republic of China between 1950 and the present day.

Since science is an indirect key to political power, is it not political? The answer is again no, in that science is a key to almost anything in modern life, and hence there is no conflict about whether science should or should not be pursued. The fact that politicians, as well as lawyers, engineers, doctors, factory workers, farmers, sculptors, or burglars, use some of the products of science-based technology does not make science political except in the sloppy, broad terminology that we mentioned earlier.

To summarize, science should be maintained and supported regardless of political differences or divisions, since it is a necessary ingredient in our lives regardless of what national, international, or ideological politics will prevail in the coming decades.

Moral questions connected with politics that may arise in the pursuit of science are discussed in Chapter 10.

8
HOW MUCH SCIENCE?

I*N PREVIOUS CHAPTERS* we stated that science is a worthwhile activity which, if pursued and supported with skill and dedication, can have a great impact on our lives. Such a statement, however, is merely idle unless we also face the problem of how much science should be undertaken. Should each of us simply drop a penny or two into a charity box every year or are we asked to make serious financial sacrifices to maintain a huge scientific establishment? Will science demand only a few odd brains here and there from among the talent that is available or will it drain away much of the manpower that might be crucial in other areas of human endeavor? Furthermore, with the rapid growth of science described in Chapter 6, what can we expect in the future as far as financial and manpower resources are concerned? Even if the balance between science and other activities is reasonable today, will it perhaps get out of hand in the future?

Ultimate Limits

To begin with, let us ask: How much knowledge is there altogether for us to explore and understand? Are there any ultimate limits to the scientific knowledge we can acquire?

These are very difficult questions to answer. Furthermore, as we learn more and more about nature, our way of thinking about these questions may change. So all we can do now is list some factors that, according to our present way of thinking, have a large role in determining the answers.

In Chapter 3 we discussed two aspects of making progress in science: the conceptual aspect (paradigm changes) and the predictive aspect. Just to make the present discussion simpler, we will think of the amount of science in terms of predictive power—that is, in terms of the realm of natural phenomena that we can predict on the basis of our understanding of science.

Another question we might ask is whether the totality of the laws of nature is finite or infinite. Naturally, we do not know. But in practice, the answer may not matter. We shall see that there are so many other factors that may limit our ability to learn about nature that whether the body of laws we want to understand is finite or infinite could very well be a purely academic question.

Most of these other factors have to do with the possible limitations of human beings, but one is not in this category, so let us start with it.

There is a very general law of nature that states that the whole world, or any isolated and insulated part of the world, if left to its own devices, will go from a more ordered state to a more disorderly one. Here is a trivial example: Even though you may clean up your desk every Sunday, during the week, just by the natural course of actions and transactions, it will get more and more disorderly, and unless you put in some effort to arrange it again, it is highly unlikely that it will, by itself, return to an orderly state.

The second half of the previous sentence also gives a clue to a second statement—namely, that it is possible to "beat" this law of increasing disorder (the technical term for it is "the second law of thermodynamics"), but one must provide outside energy to bring about such a reversed behavior.

Virtually all the interesting, beautiful, exciting, worthwhile happenings in our lives are manifestations of such reversals of the second law, since they all involve the creation of more complex, more orderly structures. Thus all technological, social, artistic, and cultural progress of humanity is based on a growing ability to utilize the energy sources available to us. On the surface of the earth, this energy is provided by the Sun (through the creation of coal and oil deposits or through the stimulation of biological growth) or by nuclear fission. Since the energy of the Sun also originates in nuclear (fusion) reactions, we can say that we owe everything in our lives to nuclear energy.

Acquiring knowledge, information, and understanding is also a reversal of the second law of thermodynamics, since it creates order from disorder. People who work in what is called information science have explored this connection between information acquisition and the second law, and can even quantify the extent to which order increases as a result of the acquisition of a "unit" of information.

Thus, in order to explore the laws of nature, in order to learn more and more science, we must apply outside energy sources to our efforts. Is it then possible that the total amount of energy available in the universe is insufficient for the acquisition of all of scientific knowledge? If so, our efforts to learn about the universe would come to a halt because of circumstances beyond our control.

Even this constraint, however, may be purely academic, even if it exists, compared to some constraints that are more down to earth and originate with us human beings.

For one thing, we as the human race might not have enough time to learn everything there is to be known about nature. Species of living beings seem to have a finite lifetime—they evolve, flourish, and then become extinct. Whether the human species will follow this pattern is not known. If indeed we have only a limited amount of time to explore the mysteries of the universe, we might not make it entirely by the time it is time for us, the human species, to die.

Furthermore, even during this limited period, only a limited number of human beings can pursue science. Nowadays, decreasing or limiting this number is fashionable. A few decades from now, once space colonization on a large scale has become a reality, the fashion may change, but, in any case, there will always be some limit on the potential scientific manpower available. We cannot expect everyone to be able to pursue science or be interested in doing so. Even if, by eugenics, we succeed in growing a mentally superior human race, other human activities will compete for people and attention.

There is, however, the possibility of building electronic brains that can do science for us, and thus we might be able to extend the effective number of scientists. This extension, however, also has some limits, in terms of technology and material, which will be mentioned shortly.

Not only are there a limited number of scientists available for exploring the universe, but it also seems as if science is getting increasingly more difficult for the human brain. There may be good reasons why this is so. As mentioned in Chapter 1, when modern science began, it concentrated on phenomena that were accessible to direct human observation and experimentation. Later we increasingly turned to laws of nature that required the study of occurrences that were quite distant from our everyday experience and senses. Such investigations required an increasingly more complex technology of instruments and machines that could create the appropriate conditions on command and then could measure the effects in which we are interested. However, such investigations possibly also require an extension of our brainpower, a more novel, unconventional, complex, unnatural mode of thinking and understanding that we may not be up to. There appears to be some sign of this already, since the laws of the so-called elementary particles have largely eluded us so far, despite research efforts of unprecedented magnitude.

The increasing remoteness of the phenomena we are investigating also has necessitated an increasing complexity and size of instrumentation, and this in itself may limit our scientific explorations in two ways. First, it is possible that our technological inventiveness will fail when it comes to devising more and more sophisticated gadgetry for scientific research. Second, even if our inventiveness continues to function, material limitations may preclude the construction of all the machinery needed. The unit cost of science is already increasing at a precipitous rate in many fields of research. The greatest scientific discoveries in the 19th century were done on budgets of a few hundred or a few thousand dollars, while today in the areas of "big science" (elementary-particle physics, astronomy, space physics) an experiment may cost millions of dollars. This inflation in the cost of science, over and above the ordinary economic inflation of all commodities, seems to be inherent in the remoteness of the phenomena we are interested in, and hence is likely to stay with us and possibly limit us eventually.

The problem of organization and management of science is another kind of constraint in future scientific exploration. As science involves more and more people and resources, it runs into the bureaucratization that all of us are so painfully familiar with nowadays. An increasingly larger number of talented scientists are robbed of their creativity, energy, and time by having to interact with the comatose bureaucracy that pervades the institutions that house, support, supervise, regulate, utilize, investigate, and administer scientific activities. Unless our ability to operate efficiently large-scale, productive activities evolves at least as rapidly as the size of those activities, bureaucracy may eventually kill the very activities it is designed to support. The successful solution of the problem of large-scale management depends also on public attitudes. How people answer such questions as "To what extent can I trust others?" or "Do I want to prevent all misuse of funds even at the cost of obstructing the productive use of them?" or "Do I want egalitarianism even at the cost of making everyone worse off?" may very well determine the difference between success and failure in managing large-scale undertakings. As such public attitudes fluctuate, eventual success in our attempt to learn such management is also uncertain.

Finally, there are the questions of motivation, will power, and interest. Fascination with science and the world around us is not necessarily a universal trait, and in fact throughout human history the degree of science-mindedness of peoples, civilizations, cultures, and ages has varied considerably. It is quite possible that most men will eventually lose their interest in doing science, and will turn to other types of aspirations and undertakings. Without a positive answer to the *why* of doing science, there will be no science.

How Much Money for Input?

All this sounds rather far-fetched, and in fact it is. So perhaps it is time now to turn back to present-day reality and to ask much more modest but, in the short run, more pressing questions about the amount of science that should be pursued.

Although we cannot predict, or plan reliably for, how much science we will get as an output of scientific activities, we can assume that the output is likely to be larger if the input is larger, and so it is reasonable to discuss the amount of input that we want to devote to science.

Input, as we have seen, consists not only of money but also of other tangible and intangible ingredients. Among these, manpower will be discussed in greater detail in Chapter 9, some detailed aspects of money in Chapter 11, and some of the intangible aspects in Chapter 10. Here, therefore, we shall concentrate on some overall considerations about the financial support of science.

In the "good old days," when the research of a top scientist took only an amount equivalent to a small fraction of his salary, or even to his whole salary, funds for financing science could be acquired from "charitable" private sources, from the petty cash boxes of educational institutions, or from small, obscure governmental bureaus. In some cases, a scientist could cover the expenses by his own private means. But those days are gone forever. Virtually all scientific research nowadays is financed either from governmental funds especially appropriated for that purpose, or by private industry with an equally well-identified research budget. (Some work is supported by private charitable foundations, but quantitatively it is relatively negligible.) Thus, the question "How Much Science?" has become a public issue and is not simply something that can be relegated to the whims of a few scientists and their affluent friends.

Probably the best way to discuss the amount of financial input for science is in terms of a percentage of the gross national product (GNP—the total value of the goods and services produced in a country in a year). This index is by no means perfect since, at least in the opinion of some people, it fails to distinguish between "useful" and "useless" goods and because it cannot include aspects of the "quality of life." Yet the GNP is useful as a purely economic measure that indicates the level of material production in as value-free a way as possible. The GNP is a much better measure to use than a government's budget, since the fraction of the former that constitutes the latter varies drastically from country to country. In an economic system of nongovernmental enterprise the fraction is relatively low, while in an economic system of governmental monopoly the percentage can be extremely high. If, therefore, we are interested in the relative size of the scientific effort as compared to the overall activity of the country, the GNP is a much better base to use.

It would be lovely if our state of knowledge and understanding were so advanced that, based on economics, cultural history, sociology, political science, psychology, and other ingredients of the science of science, we could calculate an "optimum" figure for the percentage of the GNP to be spent on science in a given country. Unfortunately, we are very far from being able to do that. Instead, we must rely, at present, on guesswork based on historical precedents and statistical information regarding those countries that seem to have been successful in developing and utilizing their science.

The statistics readily available deal primarily with research and development (R&D)—basic and applied scientific research as well as technological development and the construction of prototypes. Present-day figures for R&D expenditures as a fraction of GNP in countries where science and technology are well developed (Japan, Australia, and most countries of Europe and North America) range from 1.5% to 3.5%.

This level of R&D expenditure is a relatively new phenomenon. Before World War II, the corresponding figures were about one-tenth as much—that is, in the range of 0.1% to 0.5%. Between the war and the early 1960s, there was a rather rapid growth in the percentage of GNP spent on R&D, and since during that period the GNP's themselves grew rather rapidly, the actual increase in R&D spending each year was considerable. That rapid expansion created a psychological climate for scientists and technologists in which almost everything seemed possible and virtually every new idea received sympathetic support.

At least in the United States, this period came to an abrupt end in the 1960s. The percentage of GNP spent on R&D suddenly leveled off, and in fact even declined slightly. With the constant increase in the cost of science (an increase faster than inflation), with new scientific and technological manpower still pouring out of educational institutions, and with many research projects counting on continued long-term support for completion, tensions developed, some unemployment set in, and the morale of the scientific and technological community dropped. All this coincided with, and probably was partly caused by, a new wave of back-to-nature, anti-science and anti-technology sentiment, perhaps not large in extent but fanatic in its intensity and in the means used to gain attention.

A similar leveling-off occurred in other advanced countries in the early and mid-1970s. Thus the half-century 1935–85 has seen a jump of the percentage of GNP spent on R&D from a lower plateau of, say, 0.3% to a higher plateau of 3%.

Among the remaining three-quarters of humanity who live in the developing countries, however, the fraction of GNP spent on R&D is still in the 0.1% to 1% range. Furthermore, the total GNP's of the developing countries also are small, since their large populations are more than offset by poverty—that is, by very low GNP-per-capita figures. The combined result of these two factors is that virtually all scientific and technological research today is performed by only one-quarter of humanity.

How to Divide the Pot

The amounts mentioned thus far pertain to total R&D, including technological development. Roughly one-third (a fraction varying somewhat from country to country and from year to year) is spent solely on scientific research. Since technological development work is much more expensive per project than scientific research, it receives the lion's share of the financial resources.

Within the amount spent on scientific research, there is a division between basic research and applied research, at a ratio of about 1:2. Thus of the total amount spent on R&D in the United States in 1976, for example, about 12% was spent on basic research, about 22% on applied research, and about 66% on technological development. These figures vary somewhat from year to year.

Taking the United States again, R&D funding is divided between the federal government and private industry, and other sources contribute only a negligible fraction. About 1970 the division between the federal government and private industry was at a ratio of 2:1, but in the late 1970s the ratio was closer to 1:1. Thus, virtually all the relative decline in U.S. R&D activities in recent years was due to the sharp decrease of federal support of research. This decline was conspicuous either as a percentage of GNP or as an absolute amount in constant (uninflated) dollars. In contrast, industry's support of R&D increased considerably in constant dollars, and it did not fall even in terms of percentage of GNP. Clearly, therefore, blame for the detrimental effects of dropping R&D activities within the United States can be squarely placed on the federal government.

Funding for basic scientific research, applied scientific research, and technological development (which overall in the United States in 1976 was 12%, 22%, and 66%, respectively) differs considerably between the federal government and private industry: 16%, 24%, and 60% in the former, 4%, 21%, and 75% in the latter. To put it differently, 46% of the overall R&D was funded by private industry and 54% by the federal government; the corresponding pairs of percentages in basic research were 18% to 82%, in applied research 42% to 58%, and in technological development 51% to 49%. Thus, the role of the federal government is much stronger in basic scientific research than in the other two endeavors. This is understandable since only the largest private industries can afford to invest in activities that will bring financial returns only in the more distant future.

The foregoing figures for the United States pertain to the sources of the funding and not to where the research is performed. If we look at the latter, the picture is drastically different: 70% of total R&D expenditures is spent in private industries, 10% in universities, and 5% in other extragovernmental research centers. Only 15% of total R&D expenditures are channeled through federally administered institutions. In other words, the largest part of federal R&D funds is spent through research grants and contracts awarded to private and nonprofit organizations. This method of financing has been extremely successful in the history of American science and technology, because management of most R&D activities has been transferred out of government and into the hands of smaller, more decentralized, more flexible, much less bureaucratic organizations. One can appreciate this point by imagining what R&D activities would be like if they were operated by organizations set up on the pattern of Amtrak, the U.S. Postal System, or the U.S. Department of Health and Human Services.

Since the extent to which R&D funds are spent on military rather than civilian research is often discussed, some figures might be interesting. Practically all R&D financed by private industry is spent on civilian research, but of the reearc financed by the federal government in 1976 about 50% was for military purposes and 50% for nonmilitary purposes. In 1969 the two figures had been 53% and 47%, respectively, hardly different from 1976. There was, however, a sharp tendency in civilian research to spend less on space research and more on other research. In 1969 the ratio between the two was 1:1, while in 1976 it was 1:3.

Relative to most other national allocations, the amount of money spent on R&D is very small. In the United States in 1978, for example, the 2.2% of the GNP spent on R&D altogether (or, especially, the 0.7% spent on science) contrasts sharply with at least ten times as much (over 20%) spent on welfare programs, at least three times as much spent on medical expenses, about four times as much on military matters. In fact, the value of wasted food thrown into American garbage cans each year is about equal to the amount the United States spends on all applied scientific research. Such comparisons are perhaps specious, because it would be difficult indeed to reduce food waste and use the savings to increase research expenditures. Yet these figures do give an idea of the very small fractions of national wealth that are required for research activities. From this point of view, research seems an extremely good investment.

The distribution of research funds among various scientific disciplines is also of interest. We mentioned in Chapter 7 that the relative balance between various fields in science can be somewhat influenced by extrascientific factors, but on the whole the balances are fairly well set by the internal dynamics of the development of science itself. This is quite well reflected by figures for the United States. For example, the federal funding of basic research in life sciences, physical sciences, environmental sciences, and engineering showed a quite stable pattern between 1969 and 1976, though there was a low long-term trend in favor of life sciences and away from the physical sciences. In 1976, of the funds spent on these four areas, 41% went to life sciences, 31% to physical sciences, 16% to environmental sciences, and 12% to engineering. Apparent changes in such percentages are sometimes due not to the actual change of research fields but to an altered way of labeling research. For example, during the "war on cancer" in the United States, a considerable amount of ordinary basic biological research was labeled cancer research in order to attract more funds. Since we know very little about the basic causes of cancer, such a relabeling was neither difficult nor dishonest.

Within a given scientific discipline, there can be considerable differences between research areas in terms of the specific cost of doing research. In physics, for example, work in solid-state physics is relatively inexpensive because the instrumentation is relatively simple and relatively small. A solid-state experimental physicist can do fine research working alone (or with one assistant) in a room measuring, say, 20 x 30 feet, on a budget of, say, $100,000 a year. In contrast, an experiment in high-energy physics involves the teamwork of a dozen people, may cost $10 million and require a complex of accelerators and particle detectors measuring a mile or more in length or diameter. In such cases, the capital investment in the complex itself (in addition to the cost of the experiments performed on it) may be in excess of $100 million. The same is true for experiments in astrophysics involving satellites. Big science, as this kind of science is called, is a relatively recent phenomenon. Before World War II virtually everything was carried out by individuals or small groups using very small budgets. The changing patterns from little science to big science have had a marked influence also on the community of scientists and on the qualities needed to be successful in scientific research.

When it comes to big science and the appropriation of huge chunks of money, one needs a balanced blend of scientific and nonscientific factors in order to attract public support. When, a few years ago, the United States decided to build the National Accelerator Laboratory to house what was then the world's largest accelerator, the capital investment of some $300 million was enough to make the location of the new laboratory a political as well as scientific issue. The choice was therefore made in an appropriate way: First, using scientific criteria only, some six final sites were selected out of the dozens proposed. All six sites satisfied the scientific criteria approximately equally. Out of these six, one was selected on the ground that the Middle West had been relatively underdeveloped in terms of scientific facilities and hence deserved the new laboratory. As a result, the new institution was located just outside Chicago.

9
HOW TO CREATE AND MAINTAIN SCIENTISTS

CREATING A PRODUCTIVE and viable body of scientists is a very lengthy and demanding task—one that takes decades of time, light years of patience, and a good dose of luck. Therefore, once such a scientific infrastructure exists, it constitutes a considerable national asset which must be fostered. Scientists take care of themselves and will produce prodigiously if only they are given an appropriate framework and environment in which to work. This chapter and the next chapter elaborate on how such a community of scientists can be created and maintained in a productive state.

Is Science Elitist?

Since, however, we live in an age when charges of elitism are quite frequently hurled at people, groups, activities, and resource allocations, let us begin by investigating the extent to which science in fact is elitist.

Although the word *elitist,* like most political slogans, is imprecise and ambiguous as it is usually used, it has a multitude of precise meanings and dimensions, four of which we shall consider here.

First, an activity may be called elitist if only a few people benefit from it. In Chapter 2, we concluded that science, directly or indirectly, benefits virtually everyone, although not equally. The cause of the disparities, however, is not in science itself but in the long chain of auxiliary activities (technology, education, economics) that links science with individual human beings.

Second, science may be called elitist if its practitioners receive undue benefits, rewards, and prestige from pursuing the profession. On the whole, this is not so, though on occasions (more so ten years ago than today) pronouncements by scientists on nonscientific matters received more attention and weight than they deserved. Financially, scientists are fairly well-paid members of society but not nearly as well paid as lawyers, doctors, or managers. So on the whole, we cannot call science elitist on this count either.

Third, we may call something elitist if only a small number of people can practice it. In this sense science is elitist, but not more so than most other human undertakings that require high skill and talent. There are very few good musicians, astronauts, master carpenters, football players, or master chefs. Nevertheless, if we consider the entire group of people who are very good at doing one thing or another, this group is not so small and elitist. Fortunately most of us can claim a notable degree of accomplishment and skill in some aspect of life, and this gives each of us some of the self-confidence, pride, and satisfaction that are necessary for a contented life.

Fourth, science might be called elitist if participation in it were determined not by criteria internal to science itself, but by outside, extraneous criteria. In this respect, science is elitist on a worldwide scale since one-quarter of humanity provides virtually all scientists in the world and three-quarters of the earth's population are mere bystanders who must settle for the few crumbs that happen to fall their way from the rich set of benefits science offers.

How Many People?

Turning now to the creation of scientific manpower, we should ask the question touched upon at the beginning of Chapter 8: What is the right size for such manpower? Do we strive to make every second person a scientist, or do we talk about a few extremely rare geniuses amid a large sea of laymen?

Here also our theoretical knowledge, our understanding of the science of science cannot yet provide a definite answer. Again we must look at those countries that appear to have a productive scientific infrastructure.

We find that in such countries the number of scientists and engineers engaged in R&D constitutes between 0.1% and 0.4% of the total population. In other words, among a randomly selected group of one thousand Americans (including children, retired people, and others) we would find between one and four who are engaged in research or development in science or technology. Thus although we are very far from having every second person work in science or technology, such persons are not extremely rare.

The above percentages could have been guessed on the basis of the financial allocation for R&D. We noted in Chapter 8 that, in the advanced countries, from 1.5% to 3.5% of the GNP is spent on R&D. If we also know how much it costs, on the average, to support one scientist or engineer for one year (including his salary and research equipment), we can calculate the number of scientists and engineers we can afford to have. As it turns out, the amount needed to support one such person is about ten times the GNP per capita in advanced countries, hence the number of research scientists and engineers we can support should be a percentage of the total population that is about one-tenth the percentage of the GNP we spend on R&D. This is indeed the case in the above figures. (The situation is drastically different in the developing countries, as will be seen in Chapter 13.)

Thus, in the United States, with a population of about 220 million, about 600,000 scientists and engineers are involved in research and development. About a quarter of a million of them are scientists who work in basic or applied scientific research.

Since the size of this group increases with the population, with the GNP, and with the relative size of the R&D budget, and since losses are constantly sustained on account of people who die, retire, or drop out of research, the supply must be replenished with a steady stream of new people. Hence we shall now comment on some aspects of the education of future scientists. The science education of the population as a whole (and not only of specialists in science) is discussed in Chapter 12.

Educating Scientists

We should begin by emphasizing that becoming a scientist is an instrument of social mobility in almost all countries. Because of the objectivity and universality of science, the qualities needed to be a productive scientist depend much less on the structure of society than in the case of many other occupations. As a result, outstanding scientists can arise from any layer of the community or country. A search for, and an early identification of, prospective talent is therefore of great importance, and talented students should be encouraged to become scientists.

There are many obstacles to such identification and encouragement. The most important features of a scientific mind are inquisitiveness, an impulse to ask questions, an ability to be innovative in new concepts and ways of looking at problems, a knack for solving problems, and a capacity to generate new questions, new answers, and new connections among seemingly unrelated occurrences. Science is a wide-open subject, in which knowledge and understanding have grown at a precipitous rate, and where, as a result, the memorization of facts has only a very minor role in comparison with the other attributes listed above.

These, however, are not attributes that society can easily recognize or that it admits to be most crucial. It is not accidental that the public image of a brilliant person, a genius, a wizard is a person who wins prizes on TV quiz shows. Such shows test only the memorization of miscellaneous facts, and it is implied that a smart person is one who has crammed his mind full of such facts. Another implication of such quiz shows is that questions have one correct answer, and the "brainy guy" is the one who gets all the correct answers.

All this is completely antithetical to what is needed to become a creative scientist. Yet our schools are to a large extent accomplices to the TV shows in stressing the absorption of cut-and-dried knowledge, and grades are largely based on demonstrated excellence in reproducing such bits of knowledge in examinations.

If the situation in schools is quite bad in the advanced countries, it is a hundred times worse in the developing countries, where the early identification and encouragement of scientific interest and talent are especially difficult. Novel, inquisitive, creative minds are often turned off by incessant demands for rote memorization.

Should a champion of memorization turn to science nevertheless, he is in for a shock, since even if he has memorized all the facts that are known at the time of his graduation, a decade and a half later he will know only half of what is then known, and three decades later only a quarter. Keeping up with the memorization of scientific facts is as hopeless as it is pointless.

Another feature of science education results from the cumulative nature of science. As we saw in Chapter 3, research in science is based on work done previously, and the ways of thinking, the understanding reached previously are instrumental in subsequent progress. As a result, education in the sciences is more hierarchical than in many other fields. Science courses at the university level almost always have prerequisites; they require the successful mastery of the content of more elementary courses. In many other university subjects, however, an advanced course deals not so much with conceptually more complex matters as with more specialized material. For example, a graduate quantum mechanics course in physics has as prerequisite an upper-level modern physics course, which in turn assumes that the student has taken a lower-level introductory physics course. Without such prerequisites the quantum mechanics course will be completely incomprehensible to the student. In contrast, a graduate course in 17th-century French literature, or in the geography of Scotland, or in the anthropology of Samoa, although perhaps facilitated by an appropriate introductory course in literature, geography, or anthropology, is not nearly as insurmountable without it as an advanced physics course is without its prerequisites.

A consequence of this hierarchical structure of scientific knowledge is that it takes many years of strenuous education before a young scientist-to-be can be expected to be able to do independent research along a research frontier. A bachelor's degree in any of the sciences is quite insufficient for this, and even a master's degree is generally not sufficient. Some creative researchers have not gotten a formal advanced degree but have somehow learned things on the job, but this is both difficult and exceptional. In some European countries, formal graduate education is not offered but university training is more strenuous; even so, the resulting scientist tends to be somewhat more narrowly specialized than the product of a graduate education.

The rapid growth of science also means that education does not stop with the Ph.D. degree. On the contrary, such a degree simply prepares the scientist for a lifetime of continuing education. Scientists depend crucially on constant communication with other scientists around the world, and the absorption of new ideas, new discoveries, new methods of approach must go on incessantly. It is for this reason that acquiring the right methodology of learning is so essential.

A scientist is initially educated in a particular scientific discipline. In fact, while working on his Ph.D., he specializes even more narrowly, and his thesis research pertains to one particular problem in one particular subspecialty in a scientific discipline. For example, he might be trained in physics in graduate school, where he may specialize, say, in solid-state physics, decide to become an experimentalist, and, for his thesis, do some measurements on the dependence of the electric conductivity of iron on small impurities of other materials.

When he receives his Ph.D. degree, having just completed what is often his first piece of scientific research, he may think of himself as an expert on conductivity rather than as a physicist. A decade or two later he will undoubtedly have a different view. His thesis research will have receded into insignificance, and he may, by then, work on a completely different type of problem, in a different specialty of physics, or perhaps not even in physics at all. As new scientific disciplines arise, their initial developers are scientists from other, more traditional disciplines—just as we saw, in Chapter 6, that the new interdisciplinary area of the science of science is being evolved by people with backgrounds in history, sociology, philosophy, economics, and public administration, as well as in the natural sciences. A recent example is molecular biology, where a significant number of the most creative people came from an initial affiliation with chemistry or physics. Similarly, computer science has drawn people from mathematics, physics, philosophy, and other disciplines.

For science education to prepare a student for this kind of wide-ranging career, it must be broad, it must develop general abilities for scientific reasoning and intuition, and it must endow the student with an ability to grow with new developments. This is a tall order for any educational system, and perhaps an impossible one to fill satisfactorily.

In demanding breadth and flexibility, science also points the way in other human enterprises. As science and technology progress rapidly and affect the job structure in society, people in positions not directly related to science or technology are also confronted by constantly changing job descriptions, types of responsibility, and forms of organization. As a result, nowadays even such people should expect to have to fill types of jobs that, at the time they received their school education, did not yet exist. Consequently, even in their forties and fifties, they, too, must have the background and attitude to be able to absorb new knowledge and learn new skills.

To make such a transformation possible, their original education must have had the features of breadth, flexibility, and stress on thinking (rather than absorption of facts) that science requires for its own practitioners. Furthermore, people in their forties and fifties must have the opportunity to retrain themselves for different jobs and professions. During the four or five years it may take them to reeducate themselves, loans must be available to them and their responsibilities must be adapted in such a way so as to encourage, or at least make possible, such reeducation. Featherbedding, regressive seniority, and other practices relating to job security are extremely counterproductive in this respect. That the prospect is not impossible, however, can be seen from the situation in the scientific community, where continuous reeducation has long been practiced with impressive results.

Aging Scientists

As we saw in Chapter 6, scientists tend to make their best conceptual contributions to science very early in their careers, in their twenties and early thirties. This element by itself would make a scientist who just reached forty an old man, nostalgically reminiscing about his great days in science. This, however, is by no means the only element, since generating new conceptual ideas is not the only way to contribute to science. There are other ways in which experience counts more. For example, the art of devising new experimental techniques, the ability to be a leader of a large team in an effort in big science, or the skill in devising good approximation procedures to the solution of a scientific problem are areas where an older scientist can outshine a younger one.

Even in these areas, however, there are instances when the productivity of a scientist declines, perhaps because of a decrease in his ability to concentrate over a protracted period of time, or in his capacity for keeping up with new developments in his field, or simply because of his sagging interest in and motivation for being engaged in scientific research. Such a decline may be due to a general lessening of drive and energy or to a newly developed interest in some other activity.

A universal prescription for what to do in such cases does not exist. It is, however, generally useful to offer as many creative opportunities as possible to such an aging scientist. One of these is scientific administration or the management of science, and indeed some very productive scientists have later proven to be outstanding scientific statesmen and organizers. But this is by no means the solution for everyone, and unfortunately there have been too many cases when administration was the only channel open to someone whose productivity in the sciences had declined. In many such cases the result was detrimental both to the scientist himself and to the institution in which he was an administrator.

Other channels of activity—university administration and educational activities and innovations—are open to scientists in universities. For better or worse, a university is a seemingly endless receptacle for the time and energies of some of its faculty members—in terms of faculty committees, deanships, presidencies, and the like. Some older scientists have also found a challenging second career in curriculum reforms, innovations in teaching methods and equipment, and similar education-related matters.

In private industry the administrative channel is not only open but in fact acts like a powerful magnet. It seems that more often than not, a productive scientist working in a private industrial laboratory must strenuously fend off company attempts to switch him into management.

Another kind of channel for creative activities is international scientific cooperation with developing countries. The building of science in those countries is an extremely difficult and challenging task which requires, among many other factors, active cooperation from the worldwide scientific community.

The new discipline of the science of science offers yet another area in which scientists can make significant contributions. They enter this discipline with a wealth of empirical and intuitive knowledge which, if properly used, can supplement and simplify the research by outsiders to science.

Universities, National Laboratories, and Industry

Finally, to conclude this rather kaleidoscopic set of comments on the education and maintenance of scientific manpower, let us compare universities, national laboratories, and private industrial laboratories, the three major institutional frameworks for science.

All three modes have their assets and handicaps, and among them they offer an excellent variety of opportunities. Universities are optimal for little science, especially in basic research. They give a great deal of freedom to the individual, provide automatic assistants in the persons of graduate students, and of course allow research to be combined with teaching, the latter being an advantage in the eyes of some but a disadvantage in the opinion of others. Universities, however, do not allow undivided concentration on research, but burden the scientist with teaching and, to a fast-increasing extent, administration, bureaucracy, and paperwork.

National laboratories (financed by the national government but generally managed by some nongovernmental organization on a contract basis) excel in big science, have very much better auxiliary facilities for scientists than universities do, and are very conducive to interdisciplinary work (unless the laboratory is a one-purpose installation like the National Accelerator Laboratory). On the negative side, national laboratories isolate scientists from useful interaction with graduate students, and their programs are set so that a scientist's freedom to pursue his own particular scientific interests may be somewhat curtailed. The only purpose of national laboratories is scientific research, which is an asset as long as a scientist wishes to engage in such research with singleminded devotion and does not want to earn his keep by teaching. On the other hand, the single purpose can become a burden when a scientist becomes interested in sharing his energy and interests between scientific research and some of the other areas just mentioned in which a scientific background and experience are valuable. Work in national laboratories generally combines basic and applied research, and thus offers a broader set of choices than the universities, where research is largely basic.

Industrial scientific work can vary a great deal, depending on the size, diversification, policy, and personnel of the company. Some large companies have had a longstanding tradition of supporting extensive research laboratories in which some work fits into both the basic and applied categories (as the example in Chapter 1 demonstrates). In general, however, the scientific interest of a researcher must fit much more tightly into the interest of the laboratory of a private industrial company than at a national laboratory. The auxiliary facilities (shops, supplies, computers, secretarial help, draftsmen, etc.) are generally very good at private laboratories, partly because the personnel for these services are not part of the civil service system that plagues most universities.

The mobility of scientists, at least in the United States, is quite high, hence it is common for a scientist to have had experience in more than one of the three main types of organizations. For example, older scientists tend to move from national laboratories to universities because the latter offer a greater opportunity for diversification outside scientific research.

Since science is universal, scientists also have a very high degree of international mobility. At least from a purely professional point of view, most scientists could easily transfer into a different country and be productive with no more effort than if they had changed locales within their own country. This fact was of extreme importance just before and during World War II, when refugee scientists from Europe came to the United States and played a pivotal role in the development of not only the atomic bomb but also of American science as a whole.

10
WHAT KEEPS SCIENTISTS HAPPY?

THE TITLE OF THIS CHAPTER is, of course, partly facetious, and perhaps it is also irrelevant. Although we want to keep scientists productive, do we necessarily want to keep them happy also?

In Chapter 2 we listed some of the personal motivations of scientists for getting engaged in science. We identified as internal factors the esthetic satisfaction gained from understanding the laws of nature, the satisfaction of curiosity about the world around us, and the urge to convert talent into accomplishments. We also listed some personal factors related to society around us, such as the fun of competing, the pride in gaining priority in making discoveries, and the joy of obtaining peer recogntion. Finally, there is the entirely altruistic, externally oriented motivation of serving humanity.

To ascertain, however, the extent to which these motivations lead to productivity or happiness, we must find out whether or not the aspirations that underlie these motivations are fulfilled in the course of scientific work.

External Factors

Let us turn first to a survey of some of the external factors that a scientist needs for productive work.

Take, for example, time. Science is certainly not a 9–to–5, five-day-a-week type of occupation, and so scientific research competes with other obligations in a scientist's life. In a research institute the problem is easier, since (apart from administrative duties) only items of personal life participate in such a competition with science. In universities, however, a scientist is paid, at least nominally, because he teaches, and time spent on students contributes little to scientific research. Added to the time lost in teaching is the growing burden of often inane faculty committee meetings, the mounting avalanche of university bureaucracy, and the not inconsiderable time spent on managing the grants that support a scientist's research work. In the life of an academic scientist the time demands of research and the competing, often time-consuming other obligations create a tension, a constant state of rush and deadlines, that can be borne only by a healthy individual of considerable endurance. Much can be said for combining scientific research and science education, but the pursuit of this elusive goal is certainly hard on those who undertake it.

Facilities, equipment, and its repair constitute another important external factor in the life of a scientist. An experimental scientist spends a substantial part of his time on building the apparatus he uses for experiments. Even if funds are available for the construction of equipment, the process may be lengthy and involve risks and hence tension. Take, for an example, a group that spent some five years, full time, building intricate equipment for a novel type of measurement in space physics. During these years there were no publications, no esthetic gratification from discovering new laws of nature, no satisfaction of curiosity about nature. There was only hard, fairly tedious work in advanced technology. Then, when the equipment was ready to be shipped to the rocket launching pad for the journey into space, there was the additional fear of the rocket failing and five years of work being annihilated in seconds. In this particular case, however, the launching was successful, the equipment worked in space, and so the two years after launch were paradise for the scientists: The data kept pouring in automatically, day and night, and all the scientists had to do was to tear off the computer print-out from time to time and convert it into a scientific article of considerable importance and acclaim.

That communication is of great importance to scientists is discussed in some detail in Chapter 4. In a fast-moving field, such communication can assume a feverish pitch. Those who read James D. Watson's *The Double Helix* will remember the exciting account of how the various competing groups studying the structure of the DNA molecule constantly depended on information from the other groups and from the rest of the scientific community. The pursuit of communication channels is thus also an essential but tension-filled aspect of the life of a scientist.

Interaction with management and bureaucracy can consume much of a scientist's energy. Man's ability to engage in increasingly more complex and sophisticated activities is certainly not matched by his ability to organize and manage such activities rationally and efficiently. In special cases, when society decides that it really matters, efficient management can be achieved. If this were not so, we would never have gotten to the Moon. But those are exceptions. In most instances, and practically everywhere around the world, bureaucracy becomes more and more rampant, and getting something done from an organizational and managerial point of view is more and more frustrating, costly, time consuming, and laborious. One can only hope that this trend is temporary and will eventually turn around. One might argue that it must do so, because otherwise all scientists soon will be full-time bureaucrats pushing forms and dossiers from one desk to another.

Then there is the question of freedom from politics. As discussed in Chapter 7, in various countries scientists are not guaranteed freedom from politics, but are forced to spend their efforts and time on prescribed political activities. Although, as we shall see later in this chapter, some scientists do engage in politics voluntarily, that is a quite different type of activity. An externally imposed lack of personal freedom affecting scientists is definitely detrimental to scientific activities.

There are, from this point of view, two types of totalitarian regimes. The milder type does not demand explicit support, only a lack of opposition. Thus in such a situation a scientist who pledges to remain neutral (that is, apolitical) can continue his work without interruptions. As a human being, he may be bothered by the political state of his country, but his strictly scientific self is not interfered with. In fact, in some cases scientists are known to have taken flight from the political reality by plunging into scientific research at an increased pace.

The more destructive type of totalitarian system is one that does not tolerate neutrality but demands writing political tracts, attendance at party meetings, and various other forced acts of political submission. Under such circumstances, engaging in scientific research becomes very difficult indeed.

From the above list of external factors affecting the work of scientists, it is clear that happiness, in the simple sense of the word, is hardly a proper description of the atmosphere in which a scientist finds himself. The state of mind is much more complex. The following might be a more apt description of what is happening: a burning internal drive, a constant tension in time, space, and vis-à-vis the aspects of life that interact and interfere with scientific research, a steady state of excitement, with occasional periods of exhilaration, alternating with moments of discouragement.

Professional mobility is an effective safety valve for the frustrations encountered in the daily life of a scientist. When frustrations have mounted to an unbearable state, when expectations and drive are not matched by the realities of a particular environment, a scientist may move to a different position, a different institution, a different geographical environment. In moving, he has an opportunity to leave his faults behind and take his virtues with him. Similarly, he can leave his enemies behind and continue to maintain contact with friends.

It is unfortunate, therefore, that job security is valued so highly and professional mobility is correspondingly more and more constrained. In some countries, such as the Soviet Union, scientific job mobility has always been practically nonexistent, and there is anecdotal if not statistical evidence that this has had a negative effect on Soviet science. It is hardly an example another country would want to emulate.

Happiness from Within

So far, however, we have discussed mainly external factors that influence the state of happiness of scientists. The more interesting and perhaps also more influential set of factors is connected with the actual extent to which the internal motivations of scientists match the psychological and intellectual realities of scientific work. Let us now turn to those.

From the foregoing it is evident that working in science is an absorbing undertaking. The scientist gives much or even most of himself to the task. At the same time, the results are by no means guaranteed. Getting involved in science is very risky.

First of all, there is the question of talent and ability. Unfortunately, interest and motivation are not always coupled with talent, ability, and accomplishments. As we saw in Chapter 6, some scientists contribute very little or hardly at all and only a small fraction of the scientific community produces a substantial amount of new science.

If so, then some of the theoretical motivations listed in Chapter 2 may not have much of an effect in reality. Esthetic enjoyment of nonexistent discoveries, curiosity about laws of nature we do not understand, the conversion of talent into accomplishments that do not materialize, competitions that we always lose, priorities we never attain, peer recognition we do not receive, and service to humanity we are incapable of—these will hardly bring satisfaction.

And yet the situation is not as dismal as the previous paragraph would suggest. Almost all scientists contribute at least a minuscule amount to science. On the other hand, even the most brilliant scientists can easily convince themselves that their contributions are negligible if looked at from a distant perspective. The existence of objective criteria to measure scientific progress has both assets and drawbacks. These criteria enable a scientist to ascertain that he contributed at least a little, but they also point out the impossibility of one given human being contributing to an overwhelming or even substantial degree. Einstein's name may be forgotten in ten thousand years, and his insights into the laws of nature may appear trivial and negligible in comparison with what the state of knowledge and understanding will be by then.

Thus the satisfaction a scientist receives from doing science does not, and cannot, stem from any absolute assessment of his contributions. This is demonstrated clearly by the fact that while many very accomplished scientists behave insecurely and exhibit an obvious lack of inner serenity, others, with much more modest contributions on an absolute scale, are content, well balanced, and serene.

Another important facet of a scientist's aspiration and satisfaction is the witnessing of great progress in his branch of science during his lifetime. This is one of the reasons why some scientists are so impatient and why they constantly demand more and more resources and better and better facilities with which to pursue scientific research. It is useless to try to pacify them by pointing out that if resources are spent at a slower rate, and scientific research is practiced at a reduced pace, the eventual result will hardly be different, except that a given discovery will perhaps be made fifty years later. They want to see the new results now and not fifty years later. For this reason, among others, scientists cannot be trusted to determine, by themselves, the amount of resources society should spend on scientific research. Their drive and restlessness are extremely healthy forces, but they must be balanced against other, broader considerations.

Moral Responsibility?

Let us now turn to an aspect of science and scientists that is much discussed these days, although it is hardly a new issue: the moral implications of science and scientific research.

The interest of scientists in this issue is extremely variable. Some profess a great preoccupation with the moral consequences of scientific research, while others pay little explicit attention to it. Such a difference is not due to some scientists' being more moral than others, even though some simplistic analyses attempt to reduce the situation to such a naive cause.

Instead, the difference is due mainly to the scientists' conjecture about the extent to which any moral consequences of scientific activities can be controlled by scientists. At one end of the spectrum, there is the view that the pursuit of scientific research and the striving toward scientific discoveries are so ingrained in mankind that scientific research will continue regardless of the moral consequences. Proponents of this view point out that most of the consequences of science are good, so that the net effect of pursuing science is favorable. They also contend that scientific discoveries are morally neutral, that only the uses to which they are put by human beings can be classified according to moral categories. "To every man is given the key to the gates of heaven," states a Buddhist proverb; "the same key also opens the gates of hell." As a consequence, the conclusion is made that scientists have no more moral responsibility for the consequence of science than do ordinary citizens.

Those who take an opposing view tend to deny some or all of the above assertions, and claim that scientists have a very grave responsibility for seeing to it that mankind uses scientific research for good ends. The fact that it is impossible to decide unambiguously what is good and what is bad does not dampen the ardor of those who advocate that each scientist should assume such a responsibility according to his own standards of good and bad.

This brings us to the subject of scientists in politics—not external inferference by politics in science, which we discussed earlier, but the voluntary involvement of some scientists in issues that are to a large extent political.

There are a number of reasons why a scientist may divert some of his energy and time to political activities. Perhaps the most rational of these is the feeling of responsibility mentioned above. Much of the involvement of scientists in politics centers around such cosmic issues as war and peace, survival of the human race in the face of purportedly fatal pollution effects, and lofty manifestations of human rights. Relatively seldom does one see scientists involved in the daily, practical aspects of politics, such as whether a given school should be built here or there, or whether welfare benefits should be raised by a specific amount or percentage. It is easy to explain why scientists prefer cosmic issues, not only because of their general feeling of responsibility for the consequences of science, but for other reasons also, some of which give us an insight into the degree to which scientists are happy.

First of all, scientists are strongly tempted to get involved in politics because they have relatively easy access to the public media. That scientists are regarded as very intelligent people involved in difficult, esoteric undertakings leads reporters and others to consider newsworthy the statements of famous scientists on almost any subject, whether or not they are based on actual expertise. Few people can resist the temptation of a ready-made platform from which to vent their ideas.

More important, however, scientists are motivated to get involved in cosmic issues of politics because they realize, sooner or later, that their contributions to science are minuscule, even if they are among the most distinguished researchers of their time. After twenty years of strenuous research, and with the onslaught of middle age, some scientists may feel the need to make more momentous contributions to the world. The broad issues of politics and public life offer an apparently ideal arena for such contributions. The issues are crucial and are enveloped with nobility, and scientists also tend to believe that contributions to these problems do not require systematic training and preparation on any significant scale, but that a great mind, and a heart in the right place, can do the job.

Looking at the situation, however, from a somewhat more detached point of view, one finds that scientists generally are ill equipped to make significant contributions to such problems. They are generally contemptuous about the value and accomplishments of the social sciences, and hence tend to be impatient with techniques developed for dealing with political problems. They are overly influenced by the method of controlled experiments and theoretical models and try to apply the scientific method to human and political problems. They place an excessive premium on simplicity, which indeed is the hallmark of a good solution to a scientific problem but is seldom the key to solving social and political problems. They are also swayed by the existence of a unique solution to a problem in the natural sciences and hence do not easily comprehend the multidimensional thinking and pluralistic solutions that characterize human conflicts.

This somewhat negative assessment does not mean, however, that scientists could not make important contributions to public debate on broad issues. Many of these issues have scientific components, so that information from a strictly scientific point of view is valuable. As we mentioned earlier, science does not provide value judgments for such a debate. All science can do is to delineate the possible alternatives, to state what is or is not possible according to the laws of nature. This considerable service can be performed only by scientists.

Unfortunately, only rarely do scientists, when called upon to participate in public debate, confine their utterance to stating the scientific facts. In most cases, what one gets from a scientist is a confusing mixture of scientific statements, semiscientific conjectures, personal opinions, and wishful thinking. (See Appendix D.)

That this is so is not altogether the fault of scientists. The scientific ingredients of a public issue seldom consist only of well explored, generally agreed-upon, objective scientific facts. In addition to these, questions arise to which the scientific answer is not yet known, or not known precisely, or not agreed upon unambiguously. As an example, scientists can place, on the basis of already available knowledge, limits on the intensity of radiation that is harmful to a human being. For those, however, who have a deep psychological fear of radiation, such a procedure will not be enough. They may claim (and possibly rightly so) that there may be long-term effects of radiation which we have not had the opportunity to discover, or that there may be effects of radiation which are so small that we have not been able to see them in the relatively limited samples we have thoroughly investigated, but which may add up to a significant amount in the whole population of the earth. Disguised in this seemingly scientific dilemma there is an underlying value judgment: In the face of unknown dangers or risks, should one behave conservatively or progressively? Science cannot answer such a question.

A number of mechanisms have been suggested to create a better separation between the scientific and nonscientific components of a public debate—among them the science court discussed in Chapter 12.

Perhaps we have strayed too far from our original question of how to make scientists happy, but the various aspects of a scientist's concern touched upon do illustrate the contention that happiness is not a simple, one-dimensional state of mind. There is indeed much that can be done externally to help scientists be more productive and possibly more content, but much that determines the state of mind of a scientist originates from within, from his view of the world, of himself, and of the purpose that he sees embodied in the universe. Such existential, teleological, and ontological questions certainly do not appear in the description of the scientific method and in fact probably do not affect science very much in the long run. They play a significant role, however, in the lives of individual scientists, and hence must be considered when we try to find ways of keeping scientists happy.

11
HOW TO GIVE OUT MONEY FOR SCIENCE

IN *CHAPTER* 8 we discussed the overall amount of science that is being supported, or should or can be supported, and we indicated the overall amount of financial resources that are devoted to the pursuit of science. Perhaps even more important, however, from the point of view of the prospering of science, is the method by which this overall amount is allocated, distributed, and spent.

In order to illustrate the advantages and drawbacks of various possible ways of deciding how to divide a certain amount of money among researchers in a given field, let us outline two extreme methods that represent the opposing ends of a spectrum of possibilities. Neither is used in practice in such a pure form, but they do constitute important ingredients in how money is actually allocated.

Dribbling Egalitarianism or Reaching Down for Merit?

We shall call the first method the dribbling down egalitarian system (DDES). Let us assume that a country's one and only science funding agency decides to spend one hundred million pesos (the country's currency) on scientific research in a given year. Using the DDES, it will then meet with the five research organizations of the country (called A, B, C, D, and E) and announce that each of these agencies will be awarded 18 million pesos for the coming year, with 10 million pesos being retained for the administrative expenses of the funding agency. The head of organization A then calls a meeting of its four divisions and notifies them that for the coming year each will receive 4 million pesos, with 2 million pesos being retained for the central administrative costs of organization A. Each of these subdivisions, in turn, will award one million to each of its three research areas and retain one million for subdivisional administrative expenses. A given research area, acting through its director, will give 200,000 pesos to each of its four research groups, with the remaining 200,000 pesos going to the administration of the central office of the research area. Within each research group, finally, the money is portioned out equally among the scientific staff, except that a certain fraction is retained for running the bureaucracy of the research group.

Under the other model, the reaching down merit system (RDMS), starting again with 100 million pesos, the RDMS science-funding agency would advertise throughout the country that it intends to award sums of money for scientific research and that it is ready to accept research proposals for individual scientists or groups of scientists. Having received, say, 1,457 such proposals, asking, altogether, for 168 million pesos, the agency would proceed to evaluate each proposal according to its absolute and relative scientific merit and then divide the available 100 million pesos among a certain number of applicants.

What are the pros and cons of these two radically different systems? The DDES can boast of being egalitarian, which in today's atmosphere (in which professions of egalitarianism are tantamount to declarations of virtue) counts as an asset. The DDES is also easy and smooth to run: No agonizing decisions need to be made, no fights among organizations, subdivisions, research areas, or research groups are likely to take place. The DDES also avoids the difficult process of comparing different fields of research, different organizations, or different approaches to research with one another. Funding in this system is quite predictable, and once the overall sum to be spent on science is known, each researcher is assured support of known magnitude, regardless of whether or not his work during the past year happened to yield spectacular results. Continuity of funding is built into the system. New, fledgling research organizations, groups, or individuals are given an opportunity to establish themselves, without having to compete, right from the start, with much better established, more experienced, and more prestigious colleagues.

The main drawback of the DDES is that it completely ignores quality, merit, and past performance. There is no need to look at the record of any researcher or research organization: Once these become a recognized part of the group among which the money is to be distributed, they are equal members with no ranking and no individual priorities. If, however, a researcher, group, or project is not included because of formal, bureaucratic rules and regulations, there is no way to correct the omission through an appeal to the importance or excellence of the new group or project. The DDES greatly favors the *status quo*.

The RDMS has virtues that are almost exactly opposite to those of the DDES. It is highly flexible, it fully recognizes excellence and merit, it promotes change by treating new people and new ideas on an equal footing with older ones, it eliminates the bureaucracies which handle the money in the DDES, and it bolsters the morale of the scientific community by operating on the principle of competition based on scientific merit.

The RDMS also has drawbacks. It cannot work properly unless there exists an elaborate system of evaluating the research proposals, assessing the merits of various projects, gauging the past performance of scientists and research groups, and determining the ability of the applicants to perform the research they propose. The RDMS can also be hard on new researchers and new research groups, since in order for a group or researcher to "get going" in research it often needs a temporary subsidy, based more on promise than on actual past performance. Connected with this, the RDMS in its pure form does not provide institutional support except through specific research projects, and hence a new institution might find it difficult to acquire funds for the general infrastructure needed to create the conditions under which good research can be performed.

In practice, therefore, one should somehow combine the two types of funding procedures and either work out a compromise system with ingreditents from each method or decide to provide part of the overall funds through one method and another part through the other method.

This is, in fact, what has been happening in the United States in recent decades. The DDES type of institutional funding has been provided to various research laboratories as well as to universities, although merit has not been completely ignored in such funding. Parallel to it, a number of science funding agencies have maintained extensive programs of individual research grants, given to single scientists or to small groups, on the basis of research proposals submitted. Governmental research funds, in the form of contracts, are also awarded to research groups in private industry for such specific research proposals. This hybrid system of providing money for science has worked extremely well and has been a very important factor in the scientific prominence the United States has enjoyed in the worldwide scientific arena.

In other countries, however, the balance between these two methods of funding has not been optimal. For example, in the Soviet Union virtually all funds are distributed by the DDES method, which, however, is modified partly by some recognition of merit and partly by political considerations. The latter has encouraged a considerable amount of jockeying among scientific institutions by various means, including the nominal association of very influential scientific personalities with a large number of institutions. Because of the centralized nature of Soviet science, certain scientists can acquire, and have acquired, an enormous amount of power in deciding funding for science, to an extent far beyond what is possible in the United States, for example.

This fact struck me vividly during my visit at an institute located in a smaller city many hundreds of miles from Moscow. I asked who the director of the institute was, and was solemnly told by the local physicists that it was Professor X (and here they named an elderly, very well known Russian theoretical physicist, a member of the Academy). "That is strange," I said. "I thought Professor X was the director of *A* Laboratory" (and I mentioned one of the largest scientific establishments in the Soviet Union, just outside Moscow). "Yes, indeed," came the reply. "He is director of *A* Laboratory, but he is also our director." On closer questioning it turned out that Professor X is the director of at least half a dozen laboratories thousands of miles apart, each of which finds it advantageous, in its search for leverage, funds, and attention, to be able to boast of Professor X being its director, even if he visits each institute only one or two days a year.

The situation in most developing countries is even more extreme. Rarely, indeed, does a developing country base any program of funding on the RDMS. What makes the situation worse is the fact that in many countries funding decisions are made by people with no background in the sciences, so that the resulting research patterns have not the slightest chance of incorporating scientific merit. Chapter 13 contains more information about this kind of situation.

Criteria for Choices

As mentioned, the RDMS cannot work without an extensive and practical system of evaluation. The task consists of two parts: an evaluation of individual research proposals, and some overall decision about the extent to which various possible areas of research should be supported. The two tasks are, of course, not unconnected, but they are distinct, since two research proposals originating from completely different fields of science cannot be compared unless some judgment is made about the relative importance of these two fields quite independently of the particular merits of each proposal in its respective field.

The evaluation of work in a given field utilizes the methods of measuring science discussed in Chapter 5. In practice, the peer review mentioned there plays an important role. Since each year tens of thousands of research proposals are submitted by scientists in the United States alone, this system of peer reviewing must be an enormous one, especially since each proposal is evaluated by considerably more than one scientist. Not even counting the journal manuscripts that also reach me for refereeing, each year I get about a half a dozen research proposals to evaluate, and this number is certainly not unusual in the American scientific community. Such reviewing is performed free of charge, as a matter of public service.

How to find criteria for making choices among scientific disciplines, scientific subfields, or various approaches and directions within a given branch of science is one of the most fascinating and difficult problems in science policy. It is important to stress that even if a conceptual framework is agreed on, the actual implementation still leaves much to the individual judgment of the decision maker.

Criteria for making choices among scientific problems or disciplines are of two kinds: internal and external (or intrinsic and extrinsic). Internal criteria pertain to the particular scientific discipline or problem itself, without considering its connection with anything outside it. One might ask, for example, whether or not the particular discipline or problem area is ripe for exploitation. Many scientific problems are exciting, yet we are completely unprepared to deal with them, because they are too difficult for our present state of knowledge and understanding in that particular field of science. For example, the task of constructing a scientific theory to explain the workings of the brain of Isaac Newton and to explain biologically why he was able to contribute to science to such a large extent is certainly a fascinating task but one that is far beyond our present capabilities. Hence a research proposal requesting funds for research on this problem is not likely to get high priority, simply because it is highly unlikely to be successful at present.

The presence in a scientific field of many able and productive researchers is another internal criterion to be taken into account. No matter how important and fascinating a field or problem may appear to people outside the sciences, if it has been unable to attract a sufficient number of able and enthusiastic scientists, funding it is not likely to produce results. To be sure, there may be ways of financially encouraging the development of such scientific manpower, for instance by providing graduate fellowships to those who specialize in this field, but this is likely to be effective only on a longer time scale. There have been recent examples in the history of science in the United States when funds were made available to certain areas of research out of political or fashion-driven motivations, but such programs seem to have resulted in mediocre research produced by scientific personnel with mixed competence and drive.

Let us now turn to some external criteria for the choice of research fields or problems that connect with other fields or problems in the sciences themselves. The study of the nuclear reactions involving hydrogen and helium, for example, is of considerable interest to nuclear physicists, because such reactions involve nuclei of relatively simple structure. These reactions, however, are also of great interest to astronomers, since all stars produce their energy output with the help of such reactions. Thus, when a decision is made about funding research by nuclear physicists on such reactions, one must consider not only the potential importance of this work in nuclear physics, but also in astronomy, even though no astronomers are directly involved in the proposed research.

The previous example also illustrates the other type of extrinsic criterion, namely the connection with foreseeable technological applications—foreseeable, since in the long run every piece of scientific research that is important by internal criteria will also turn out to have important technological applications. We can be sure of this on the basis of the past history of science, in which there has not yet been an exception to this rule. There are, of course, relatively recent scientific discoveries whose technological applications have not yet been very well developed, but as time goes on, such applications seem always to be found. This may have been in the mind of the great 19th-century physicist Michael Faraday who, when someone asked what was the use of his latest discovery, replied, "Sir, what is the use of a newborn child?"

The particular application of the nuclear reactions involving hydrogen and helium is in the practical utilization of the fusion process. The already developed version of this is the hydrogen bomb, and the slow, civilian variant of it has made dramatic progress in the last two decades and is likely to become one of our main energy sources in the first half of the 21st century.

Centralism versus Pluralism

Let us now turn to another aspect of funding: the question of centralization versus pluralism, and the related problem of duplication and waste.

To an uninitiated observer with an orderly mind, it may be evident that the best way to fund science is to have one giant centralized agency that deals with all applicants for support of scientific research and that, having made its decisions in view of all the available information, awards research grants to the best set of applicants. In practice, however, such a system turns out to be very much inferior to a different and seemingly much more disorganized one, in which there are a number of agencies to which a given researcher can apply for support. Why this paradox?

First of all, as we have seen, granting support to scientists on the basis of their research proposals is not a science in itself. It is full of uncertainties, guesses, extrapolations, and risks, and therefore even if the personnel giving out the awards were perfect, decisions (both positive and negative ones) would be made which later, with hindsight, would be judged to have been mistakes. In reality, furthermore, the personnel reviewing the applications are not perfect, especially if they represent a governmental agency. Incompetence, indifference, bureaucracy, and even political pressure contribute to such imperfections. It is, therefore, quite inadvisable to place all research eggs in one basket. It is much better to have several more or less independent agencies, with different attitudes, carrying degrees of risk-taking, and separate personnel, so that if bad decisions are made, they do not affect the whole country in a centralized way.

But how about duplication, one may ask. If several separate agencies fund research, there is danger of wasting money on two similar projects.

This is not a problem in practice, for two reasons. First, the separate agencies can (and in fact do) keep in touch with each other, so they are cognizant of the awards contemplated or given out by the others. Second, and more important, avoiding duplications in all cases is a very poor objective in science policy. Since the outcome of scientific research is often uncertain, unpredictable, and unschedulable—and since important scientific discoveries are seldom accepted without some other researchers' confirming, by experiment or calculation, the veracity of that discovery—it is in fact mandatory to build into the scientific research system what may appear to be a certain amount of intentional redundancy. One should have more than one group that is competent and interested in a given research problem. This duplication is even essential in technological development work, which is much more predictable and certain than scientific research. When, for example, during World War II the Manhattan Project wanted to develop a practical way of separating the two isotopes of uranium, several groups were formed, independent of each other, with different approaches and different personnel, to work on the same objective. Some failed almost completely, others scored some successes, and finally the best could be selected, with hindsight, to serve as *the* method used in the actual production of fissionable material.

Continuity is another important element in the funding of scientific research. Research on a given problem may take many years, and financial support must come in a steady pattern. The on-again-off-again mode of funding is very detrimental. Fortunately, governmental machinery is so ponderous and has so much inertia anyway that in practice a reasonably steady funding pattern exists at least over a period of three to four years.

"Overhead," or How to Get Public Money Under False Pretenses

As mentioned earlier, the federal government in the United States to a large extent performs its scientific research extramurally, that is, through universities, institutions, laboratories, and companies that are outside the government. This system has resulted in high-quality research. There is, however, at least one aspect of such outside funding that is very negative and deserves discussion also because it is an excellent example of an "addictive" governmental program.

When, after World War II, the federal government began to support, on a large scale, scientific research in the universities, it expressed its willingness to give them overhead payments to compensate for their out-of-pocket expenses in accommodating this new research. For example, in order to pursue a certain research problem in chemistry, a hood to vent noxious fumes may have to be installed in a university's chemistry building, which then can be financed through this overhead payment. Similarly, added janitorial service, increased administrative costs, or improved library holdings could also be conceived as expenses due to the research commitment that was assumed.

As time went on, however, this apparently reasonable system steadily deteriorated until by now it is simply a way for universities brazenly to acquire covert federal support for higher education. Whether or not this objective is meritorious can of course be debated, but the practice of getting this support out of the funds earmarked for scientific research cannot be condoned.

How did things deteriorate? First, twenty to thirty years of steady overhead payments have given state legislatures and boards of trustees the impression that scientific research performed at universities is an extra activity that universities take on, very generously, to please the federal government, and hence the universities must be compensated for the costs, whether direct or indirect. These legislators or trustees no longer recognize that a university without research would not be a university at all. They see the university only as a transmitter of knowledge, not realizing that such a function is intimately connected with generating knowledge, the other traditional and organic function of a university. This crippled view of a university results in some insultingly pedestrian ways of evaluating the efficiency of a university—for example, by the number of student contact hours per faculty member.

But this conceptual effect of overhead payments is not the

only detrimental aspect. Because of the bureaucratic complexity of calculating such overhead payments, and because of the federal government's laxness and timidity in cracking down on practices which, by the letter of the law, are permissible but which violate the spirit of the interaction between government and university, the actual amount that is returned by the university to the research group that generates the overhead is only a small fraction of this overhead payment. The rest goes, indirectly, toward the support of other, nonscientific faculty and, even worse, of the mushrooming of the university bureaucracy. It is not at all an exaggeration to say that if overhead payments on research were stopped (and at the same time the research were also stopped), virtually all universities would go bankrupt immediately, since their daily operations, completely apart from sponsored research, have become heavily dependent on this clandestine federal support to higher education. That is why this practice is also so addictive: It can no longer be stopped without heavy withdrawal symptoms. Yet the system must be drastically reformed before it does additional damage on both a conceptual and practical level.

Private Industry

That part of research in private industry which is performed through the use of federal funds is financed more or less as described above: A proposal is prepared, and if accepted by the government, a contract is written between the government and the industry for the performance of such research.

Private companies themselves, however, also finance a substantial amount of research and development. In general, the larger the company, the more likely it is that it can afford, and is interested in, longer-term research and development activities. Small companies tend to worry only about some developments in product engineering pertaining to their immediate commercial interests. One of the most conspicuous examples of large-scale, industry-maintained laboratories performing everything from basic scientific research to short-term technological development is the Bell Laboratories of the Bell

Telephone Company, which, for instance, was a main pioneer in the solid-state physics of semiconductors and the subsequent development of the transistor.

There is a practice of large, multinational companies that deserves notice because, although very sensible by short-term criteria, it ignores certain longer-range considerations. Such companies, with many subsidiaries in developing countries around the world, almost always concentrate their research and development activities in a few laboratories located in the mother country. This practice is justified by the claim (which is probably true as far as it goes) that having many small research laboratories spread all over the world would be inefficient, could not attract personnel of the highest quality, and would suffer from lack of communication with the various components of the company. At the same time, however, the lack of local research and development activity on the part of such subsidiaries deprives the developing countries of an extremely valuable channel for the transfer of science and technology. As we shall see in Chapter 13, an important shortcoming of these countries is their inability to connect technology with production, and hence the presence of industry-sponsored research and development aimed at direct utilization in the subsidiary would contribute significantly to a country's scientific and technological infrastructure. It would also help to improve the public image of multinational companies in the Third World and hence would reduce the growing antagonism toward such companies.

12
HOW TO PROVIDE FRIENDS FOR SCIENCE AND SCIENCE FOR YOUR FRIENDS

T*WO OF THE THREE* broad areas of motivation for pursuing science discussed in Chapter 2 had to do with the relationship between science and the nonscientist. Both science as a human aspiration and science as an influence on man's view of the world are closely related to how a nonscientist views science.

The average citizen seems to regard science with a mixture of attitudes and emotions. Wonder, incomprehension, fear, gratitude, and respect are only a few of the elements of this mixture.

The layman faces four major obstacles (or apparent obstacles) in trying to understand the *content* of science. First, science is internally connected, so that the understanding of certain more advanced concepts and topics depends crucially on having understood some other, perhaps more elementary ideas on which the former are based. In university terminology, science has strict prerequisites. Once someone gets lost in a science course, he will not be likely to recover unless he masters the particular idea or topic on which he got lost. Many laymen who profess incomprehension of science blame it on having gotten lost during their high school years. Such internal connectedness generally does not hold for other areas of inquiry—at least not so strictly.

Second, science cannot be mastered through the time-honored custom of memorization, which works reasonably well in many other scholarly disciplines. In science, progress is made through understanding, and once that is achieved, science becomes so simple that hardly any memorization is necessary.

Third, science has its own terminology and language which sounds frightening to the uninitiated. Even many people who are used to learning and using many languages (for example, West Europeans) are terrified by the prospect of having to learn the language of science—that is, scientific terminology.

Fourth, and somewhat connected with the third, laymen often confuse science itself with the use of mathematics in science (to be discussed presently), and so one can hear pathetic statements like, "I cannot understand physics because I don't know math." To be sure, the knowledge of fairly sophisticated mathematics is essential for some scientists in some areas of science. Yet for a layman to grasp many of the important ad beautiful ideas of science, no more mathematics is needed than what one learns in grammar school.

Werner Heisenberg makes a very clear distinction between mathematical formalism and physical ideas when he says (in *Physics and Beyond*), "The mathematical framework of relativity theory caused me no difficulties, but that did not necessarily mean that I had understood why a moving observer means something different by time than the observer at rest." To Heisenberg, unlike the layman, the mathematics came easier than a truly deep understanding of the physical ideas.

Perhaps the following dissection of scientific problem-solving will help dispel these superstitions about why laymen have difficulty understanding science.

How to Solve a Scientific Problem

In solving a scientific problem a three-step process is followed. First, the problem must be understood conceptually and qualitatively. We must be able to list the factors that will contribute to the solution of the problem, the scientific knowledge that will be pertinent to it, and the existing or hypothesized laws of nature that will affect it. Together with this conceptual and qualitative analysis, we must also make a rough estimate of the quantitative aspects of the problem.

Second, a more rigorous mathematical treatment must be applied to the problem so we can calculate its exact quantitative aspects.

Finally, having formulated the mathematical framework that can yield a quantitative result, we must make numerical calculations in order to get the actual numerical results.

An illustration of this threefold procedure is a problem which, though not a matter of science only, has a number of scientific ingredients: Can we supply Los Angeles with fresh water by towing icebergs from the Arctic into Los Angeles harbor and melting them there?

Step one is the conceptual and qualitative stage and a quantitative estimate. We conclude that, indeed, large icebergs are available in the Arctic; they move around in the ocean, so towing may be a possibility; Los Angeles is warm, so the icebergs would melt. If we assume towing at one mile per hour (which could be shown to be reasonable through another estimate involving the force needed to tow large icebergs), it would take something like six months for an iceberg to be towed from the Arctic to Los Angeles during which time, it can again be shown, the iceberg would not melt significantly. Using a volume of 2 km. x 1/2 km. x 1/2 km. for each iceberg and 200 gallons per day for each of the 5 million inhabitants of Los Angeles, one concludes that one such iceberg every three months would do.

Note that none of the above considerations require sophisticated knowledge of physics or other than grade school mathematics.

Having decided that the scheme will probably work and is therefore worth exploring in detail, we would set up the detailed mathematical formalism of calculating the force needed to tow, the rate of melting of the iceberg, the effect of a parked iceberg on the other uses of Los Angeles harbor, etc. This stage requires a considerable amount of technical knowledge of various sciences, together with a sound knowledge of higher mathematics.

Having set up all the mathematical equations, what remains is to solve them, and this, more often than not, is done on a computer.

Not all scientific problems utilize all three steps extensively, but the basic elements are usually present. The first step, in particular, is always there. This is also the step that is most accessible to the layman: It requires only the qualitative understanding of the laws of nature, and a very elementary mathematics not going beyond what is used by everyone when shopping in a supermarket.

Science and the Layman

So far we have discussed how a layman can have access to the *content* of science. Indeed, he can appreciate the ideas and viewpoints of science, and can even participate in the problem-solving aspect of science, at least in the first step, or, in simple situations when the other two steps are not significant, even entirely. There are many everyday situations when such a use of scientific thinking can contribute significantly to the resolution of practical problems.

But science also has more to offer to the layman: its methodology, its attitudes, its approach. The layman can benefit from knowing how science is done, how new ideas are born, how decisions are made about who is right and who is wrong. At the same time, science can also benefit from public understanding of its aims, methods, and needs. This aspect of science is seldom communicated to the layman, and in fact was an important motivation for my writing this book.

Indeed, the people and media who work on the popularization of science greatly neglect this second message of science.

Newspaper coverage of science, programs on television, popular lectures, science museums, and other channels of communication deal almost exclusively with the content of science, with the latest discoveries, with the laws scientists have established, and with the demonstrations of intriguing natural phenomena. This is a laudable objective, and in fact much could and should be done to improve its spotty quality. But in addition, considerable attention should be directed to the popular exposition of the aims, methods, and actual operations of science and scientists.

Without this, science will remain, in the eyes of laymen, a gee-whiz kind of operation performed by some seemingly intelligent people who use mysterious and incomprehensible methods. This magicianlike image of the scientist does not, in the long run, serve the cause of science, since, depending on the momentary public sentiment, the awe for magicians can easily turn into considering them as frauds.

An understanding of aspects of science other than its content is also important in public decision-making, where issues with scientific ingredients arise all the time. For example, if having significant scientific background is defined as having at least a master's degree in one of the natural sciences (an extremely low standard, indeed), the number of people with such qualification in the United States Congress is 1%, at most. The staffs of the various congressional committees consist almost always of generalists (people who know practically nothing about many things), and scientific background among them is also extremely rare. We must, therefore, educate the congressmen and their staffs, even if they do not have a background in the sciences, to understand science not only in terms of the elementary technical details needed for a certain issue, but also in a general sense, from the point of view of the outlook, approaches, aims, methods, operations, and needs of science and scientists.

An interesting and fairly recent suggestion of how to handle the scientific aspects of a public issue is the format of a science court. In it, two scientists would argue the two sides of the scientific aspects of an issue, just as two lawyers serve as advocates of two points of view in court. Although in the science court only scientific arguments would be allowed, one hopes that the emotional involvement of the two advocates with different aspects of the issue would spur them to give an eloquent and expert exposition of the scientific arguments on their respective sides of the issue. The jury in this science court also would consist of scientists. The outcome of this process would be a firm conclusion concerning the scientific aspects of the issue, thus confining further discussion to the other, nonscientific components. It is too early to tell whether or not this proposal will work in practice. In a recent controversy, involving a power transmission line in New Hampshire, the side that obviously was weak in scientific arguments simply refused to submit the case to a science court, but such *a priori* refusal may not take place in cases when the two sides are better balanced.

Foes of Science

Since this chapter is on how to provide friends for science, we might ask which of the motivations for doing science is likely to be most powerful in making friends for science. Scientists generally assume that the only way to sell science to the public is through the material benefits science-based technology brings. As mentioned in Chapter 2, my own experience suggests otherwise. In my frequent chance encounters, during travels in and outside the United States, with nonscientists of various social classes, educational backgrounds, ages, sexes, and nationalities, the dominant attitude I find toward science and scientists is not respect for the generating of material benefits, but enthusiasm about exploring the mysteries of nature. People seldom ask me questions about technology, but instead about some recent scientific discovery they read or heard about in some popular medium and about which they want to learn more. (Similarly, public support for sports is not based on the economic benefits brought about by the expansion of manufacturing plants that make baseball or cricket bats, basketballs, or soccer footballs.) Science, therefore, should do more to exploit its appeal to the curiosity and inquisitiveness of people generally.

Yet, at the same time that there is a fascination with and abstract admiration for science among laymen, there is also an element of anti-science feeling. It consists of claims that technology (which many people cannot distinguish from science) is bad for our bodies and souls—alienation from nature, wars, pollution, wasteful consumption, depersonalization, overconcentration of political power, and suppression of spirituality are some of the slogans used in these critiques—and that mankind is not ready for access to too much scientific knowledge in the absence of similar progress in nonscientific areas of knowledge and maturity. In other words, it is claimed that we are not ready for more scientific knowledge until we have matured considerably in social, political, and moral ways.

This phenomenon is not new. Throughout the history of science, there have been claims—the exact forms of which have changed from time to time—that science is bad for body and soul. Science has been blamed for all sorts of ills of mankind, and halting the progress of acquiring scientific knowledge and understanding has been proposed many times as a cure for such ills. For example, various religious authorities and organizations have often regarded science and scientists in a negative light, and the image of the scientist as a sorcerer emerges in the literature of many cultures.

Although, in every case, such attacks on science have failed, because they turned out to be narrow minded, claustrophobic, one-sided, and unrealistic, it is difficult to predict whether this will be true in the future. My own feeling, however, is that if attacks on science succeed, mankind as a whole will die. Whether for better or worse, mankind, some time ago, committed itself, both psychologically and materially, to a course of development based on a continued search for knowledge and understanding. A world that is characterized by no-growth in the quest for scientific knowledge is likely to become so regimented in order to suppress this quest, and so problem-ridden because of the absence of new knowledge, that it will find it difficult to survive.

It is possible, however, that such antiscientific movements will take over some parts of the world. If so, these countries or civilizations will go under, and will be replaced as leaders of mankind, by others that are more dynamic, more confident, less fear-ridden, and more strongly aspiring. There are already signs that some of the formerly leading countries are declining because of such lack of confidence, and other countries, only recently undeveloped, are rising meteorically. For example, the United States' share of the world's industrial and scientific production has declined sharply in recent years, while countries like Brazil have been increasing their share very rapidly. If such reversals take place, they will simply follow the countless historical examples of similar changes of leadership throughout the last five thousand years. (More about this in the next chapter.)

Science and the Arts

Finally, let us analyze the relationship between science and art, which may also help in promoting the link between science and the public. After all, many people are more familiar with the accomplishments of art since they themselves can explore art as amateur participants. In contrast, contact with science on an amateur basis is rare and difficult. If, therefore, the similarities between artists and scientists become generally understood, perhaps science also will cease to be a mysterious, esoteric undertaking.

There are indeed many similarities between scientists and artists. In their motivations, first of all, they share the urge toward creativity. They also share the esthetic satisfaction derived from their work. As indicated in Chapter 2, the feeling of beauty in the laws of nature is an important internal motivation for scientists. Artists and scientists also share a feeling of contributing to humanity, though in different areas. A great painting or piece of music is as much a gift to humanity as the theory of relativity.

In approaching their respective fields, artists and scientists share the element of intuition. In discussing the scientific method in Chapter 3, we saw how, contrary to simplistic popular views, intuition plays an enormously significant role in scientific research, at least as important a role as the much-heralded logic.

These similarities between science and art are indeed fundamental and basic, so much so that these activities could very well be considered as two branches of the same thing. And yet, to many people they appear very different. This may be so because when we go from general and basic features to specific details, there are in fact many differences between science and art, and these details sometimes obscure the overarching similarities.

On the motivational side, for example, scientists and artists differ in their striving for immortality. A scientist contributes to an overall, continuing structure of scientific knowledge, hence he can be content that even if his contributions are small, they will in fact be incorporated into this immortal structure. In contrast, art is not cumulative, and the value of an artist's works depends very much on the tastes, likes, and dislikes of people at various times. Art explores a variety of approaches to age-old problems; thus a third-rate artist may find it difficult to believe that his contributions to art will have the same touch of immortality and permanence as the contributions of a third-rate scientist.

Artists and scientists do differ in the degree of uniqueness of their contributions. Precisely because science is cumulative, objective, universal, and collective, it is difficult for any scientist, no matter how outstanding, to claim that had he not lived, his contributions to science would never have been made. On the contrary, it is most likely that they would have been made by someone else, perhaps a few years later. Artistic creations, in contrast, are much more individualistic, hence artists are in a much better position to take pride in the individuality and uniqueness of their contributions. As the French mathematician Charles Hermite said, "We are rather servants than masters in mathematics," and this also holds for the natural sciences. In contrast, artists are much more justified in claiming that they are masters of painting, sculpture, music, or writing.

This last difference can also be stated in a slightly different way: Artists express something that was born inside themselves, while scientists simply call attention to something that already existed before the scientific discovery. This difference makes artistic communication somewhat more difficult, since something personal and subjective needs to be transmitted to others, while in science the topic of communication is something objective and hence presumably easily available to everyone.

A rather convincing "proof" of the objectivity and collectivity of science as opposed to the individualism and uniqueness of contributions in the arts is the fact that while in science the independent but simultaneous discovery of something new by two different scientists occurs very frequently, seldom in the arts have two artists created, independently of each other, the same or very similar works of art.

Another difference, already touched upon, between science and art is that problems in science change all the time, while problems in art remain basically the same. In science, therefore, the novelty is often in the problem, while the novelty in art is in the approach used to tackle the problem.

For example, in science, the apparent motion of celestial objects (celestial mechanics), which was a pioneer field in the 18th century, is nowadays practiced by only a very small group of researchers who are preoccupied mainly with computer calculations to improve the numerical accuracy of the applications of old, well-proven principles. In contrast, the research area of elementary particles, which is today one of the chief pioneer fields and did not exist at all before 1940, today employs huge communities of scientists and consumes enormous financial resources.

In the arts, however, the subjects remain the same and only viewpoints change. The musical characterization of man facing death occurs in the 17th-century compositions of Claudio Monteverdi as well as in the 20th-century works of Benjamin Britten, but they use drastically different musical techniques to get across their ideas and feelings.

As regards the sociology of the artistic and scientific communities, perhaps the largest difference is in the collectivity of science and the individualism of art. For this reason, as we saw, scientists have a powerful need and urge to communicate constantly and extensively with each other, while artists tend to be solitary individuals. Teamwork in art is rare, but it is very frequent in the sciences.

In the sciences young people often have an edge over older ones, at least in some of the conceptually most novel and original aspects. As a result, some of the greatest contributions to science were made by people in their youth. Artists, in contrast, generally mature as they grow older, and their creations tend to become more powerful and profound.

These significant and revealing differences between scientists and artists must not obscure the fundamental, general similarities between the two kinds of human endeavor. If people consider scientists more akin to artists, they may be more stimulated to understand them and to expropriate out of their ideas, attitudes, and methods that part which can also be useful and pleasurable for the person who is neither an artist nor a scientist.

All in all, science is by no means that forbidding, esoteric, and specialized field, reserved for a few insiders, that many people visualize. On the contrary, with a relatively modest effort, but with a shedding of fears, inhibitions, and specters of huge barriers, almost anyone can make a meaningful contact with science on some level and in certain aspects, thus gaining understanding, pleasure, and a broader horizon, and at the same time benefiting science by providing it with a more understanding base of support.

13
WHO IS DOING SCIENCE AROUND THE WORLD?

A*S WE REMARKED* in Chapter 3, science is universal. At present, however, countries representing one-quarter of the world's population produce about 95% of the world's science, while the remaining three-quarters of humanity is responsible for 5%. Why is there such a tremendous disparity, and should we try to do something about it?

How the Gap Came About

Modern science was born in so-called Western civilization. Previously there were from time to time, countries with some interest in science, and some areas of science (such as observational astronomy) made considerable strides. Still, it is fair to say that the total scientific knowledge up to the beginning of the scientific revolution that took place in Western countries was negligible, and the smallness of this progress was due mainly to the absence of the scientific method.

So when modern science began to evolve some four hundred years ago in Western Europe, its progress was first slow and confined to Europe. Soon, however, it picked up momentum and became not only a supplier of knowledge to underpin technology but also a potent cultural force that changed the thinking about the world of those who came under its influence. This aspect of science was discussed to some extent in Chapter 2.

By the 20th century, science drastically altered Western society. It brought with it the industrial revolution which eventually offered unprecedented material affluence to many of the people it affected, and was also responsible for a corresponding enriching of their intellectual, spiritual, and emotional lives. Personal aspirations could be realized through professional as well as leisure activities, a rather high level of education became widespread, and society assumed an attitude of confidence, an enterprising and forward-looking spirit, and a growing pride in mankind's ability in part to control its fate.

Until quite recently, however, all this remained confined to countries in Europe, North America, and Australia. Correspondingly, an enormous gap developed between those countries and the rest of the world in the knowledge they possessed and in their material and nonmaterial standards of living. Colonization and the assumption of leadership and of the "white man's burden" seem now to be simply consequences of this colossal disparity, which previously did not exist: In material or other respects—life expectancy, incidence of diseases, knowledge about the rest of the world, technological capability, and other indicators of material and intellectual status—15th-century Europe did not differ substantially from much of the rest of the world.

This summary is not intended to offer a moral judgment on the transition of Europe from equal footing in the 15th century to a vastly dominant position in the 20th century, or on the conflicts and actions that accompanied this process. Indeed, some writers regard the rapid rise to dominant status, the methods used during the rise, and the present-day consequences of the disparity as highly objectionable. The fact remains, however, that the development of a huge gap necessarily resulted in a divided world with tensions and inequalities.

Japan was the first country outside Western civilization to catch up in this scientific revolution. Starting in the 1870s, it steadily built up its indigenous scientific and technological capabilities until, today, it is one of the leading countries in the world. In the 20th century India, China, and some other countries began to follow suit, although on a small scale, and after World War II, as many countries gained independence from their colonial rulers, the desire to participate in the scientific and technological revolution became universal. And yet, most scientific activity still takes place in the Western countries, now joined by Japan.

It is not difficult to see why this should be so. Since it took the Western countries some four hundred years to reach their present state of development, it is unlikely that other countries could reach the same stage in one-tenth of the time, even though they don't have to go it alone but have the example of and some assistance from the scientifically developed countries.

Why the process is so difficult and lengthy can be explained, in part, in purely material terms. Even if one of the poorest countries could increase its gross national product (GNP) per capita by 20% each year, instead of the actual 5% or so (indeed, a 20%-a-year growth would be unprecedented), it would still take more than twenty years to catch up in material wealth with the richest countries—assuming that the latter did not grow at all during that period.

There is a fairly direct though approximate connection between GNP per capita and the growth of science, as discussed in Chapter 6. The percentage of the GNP spent by various countries on science (*not* including technology), is, within a factor of ten, the same for almost all countries—roughly in the range between 0.1% and 1%. The cost of maintaining a scientist per year (including his research equipment) is, again within a factor of ten, also the same in most countries, and is roughly within the range of $10,000 and $100,000. Furthermore, there is a correlation between the percentage spent on science and the cost of maintaining a scientist, so that countries that spend a lower percentage also can maintain their scientists for less money. Thus, the GNP of a country determines, at least within a factor of ten or so, the number of scientists the country can maintain and hence the amount of scientific progress it could make under ideal circumstances.

This financial argument is, however, only very approximate, and variations can occur from country to country, depending on the extent to which an atmosphere that is conducive to the evolution of science can materialize.

At this point numerous nonmaterial factors enter in the actual determination of the degree to which developing countries can build up their indigenous scientific infrastructures. Unfortunately in most developing countries many of these other factors tend to be negative, and although many of these negative factors could be reduced or even eliminated through international cooperation, this has failed to materialize so as to make a major impact on most countries.

Catching Up Isn't Easy

Among these negative factors, first, there is the absence of a scientific tradition in the country, a lack of public understanding of science, and the nonexistence of a sympathetic administrative structure to aid scientists in their work. In other words, the *scientific* revolution as a cultural force has not become deeply rooted in many of the developing countries, in spite of the superficial signs of a *technological* revolution (the ubiquitous transistor radios, cars, electricity, or ballpoint pens) being very much in evidence. Thus the atmosphere in which a scientist in a developing country must work is more hostile, or at least more neutral and cold, than the environment of his counterpart in a scientifically developed country.

Second, educational facilities to produce scientific manpower are either missing or scanty, and often more intent on quantity than quality. Thus advanced education in the sciences must be obtained abroad—a process that is expensive, wasteful in manpower (since some students may not return), and also precludes the scientists back home from forming a school of young scientists around them and thus building up a scientific tradition.

Third, scientists in developing countries are isolated—a matter in which members of the worldwide scientific community could make enormous improvements without making any significant sacrifices in their own facilities. How?

Chapter 4 discussed the great variety of modes of communication that scientists use, and stressed the crucial importance of a scientist being able to keep in contact with fellow scientists. The patterns of this scientific network are largely determined by the scientific community itself. When and where to hold conferences, whom to invite to them, what journals to publish and what editorial and financial policies to pursue in connection with them, what institutions to visit on the way to scientific meetings, whom to send copies of reports, prepublication copies of research papers, or sets of data, whom to invite as a visitor to one's research institution—these are decisions made primarily by scientists themselves.

Scientists use a simple general principle for such decisions: They aim to maximize the scientific research output in the very near future. Thus, the person, the group, the institution, the country that produced best in the very near past will get the most preferential treatment, since it is judged to have the best chance to produce again tomorrow. This principle is followed not only internationally but even within a country, where leading institutions, groups, or people have the best communication facilities.

This principle represents a great disadvantage for the developing countries, which in the past have not had the opportunity to shine in science, and hence are not offered the scientific communication channels they would need to make rapid advances in science. Scientific journals reach them only sparsely and slowly because the journals, most of which are published in the scientifically advanced countries, are expensive, have to be paid for in scarce foreign currency, and travel slowly by boat unless huge additional air mail costs are paid. In many journals publication charges prevent potential authors in the developing countries from submitting their manuscripts or getting speedy publication. Scientific conferences are almost always organized in a scientifically advanced country, thus making the travel costs for the scientists from the developing countries very large. Conferences with a closed list of invitees often fail to extend invitations to scientists in developing countries who are deserving but unknown because of publication and other communication barriers. Because of geographical location, lack of personal ties, and other reasons, visits to and from scientific institutions in developing countries are much rarer than is the case in institutions located in the scientifically well developed countries. All in all, contacts between scientists in a developing country and the rest of the world of science are often lamentably weak.

There is also isolation in the area of scientific equipment. While pieces of equipment are often lacking, that is not the most crucial shortcoming. Instead, repair facilities, spare parts, and current supplies (chemicals, batteries, bulbs, etc.) constitute the real bottleneck, since most of them must be imported from abroad, through orders only laboriously constructed because of nonexistent catalogues, through shipping delays and breakage, and through domestic customs systems that are frequently a Kafkaesque nightmare. All this is done in exchange for scarce foreign currency, which causes additional delays. Technicians to effect repairs are in short supply and not well trained, so that the improvisation of spare parts out of local materials is seldom feasible.

Many of these difficulties could be considerably lessened through international cooperation and assistance. If this is so, and if the problems are so conspicuous, why has relief not yet come?

As far as the international scientific community is concerned, it has not come because most scientists are too busy in their daily preoccupation with research to pay attention to such matters as the plight of their colleagues in the developing countries. Although shortsighted, such an attitude is quite understandable. After all, a good scientist is excited, fascinated, and almost entirely obsessed with his scientific research work and feels a great compulsion to neglect everything else and create new scientific discoveries. This ethos is well reflected in the scientific community in that recognition, awards, and improved facilities are bestowed on those whose personal research work is the most successful.

This attitude is also shortsighted in the following sense: In the long run, if all of humanity rather than only one-quarter of it could be brought to the stage when it could supply active scientists, science would make much more rapid progress, and this would delight and benefit all scientists the world over. The problem with this argument is that the benefits of an increased concern for scientists in the developing countries are likely to manifest themselves on a large scale only two to three decades from now, and most scientists do not think that far ahead. On a personal level we cannot blame them—they might be dead by then.

But not only the scientific community is at fault for not providing sufficient cooperation with scientists in the developing countries. Also at fault are the many national and international assistance agencies which spend many billions of dollars a year on foreign aid activities. These organizations are also shortsighted, look mainly for results in the next few years, and in general fail to recognize the importance of the building of science as part of the development process of a country. With an eye only on short-term economic gains, the agencies contribute almost exclusively only to technological projects and fail to recognize that without a scientific base such projects are likely to fail as soon as international assistance is discontinued. Furthermore, they fail to appreciate the noneconomic justifications for doing science (outlined in some detail in Chapter 2) which play at least as important a role in development as the economic ones. They also ignore the fact that building a scientific infrastructure in a country takes many decades and hence work on it must begin promptly.

Proportions Among Those Up Front

So far we have discussed mainly why three-quarters of humanity is only marginally involved in science-making. Now let us turn to the one-quarter that is substantially involved and see how its science-making is distributed.

In getting into more quantitative details, we should decide how to measure the scientific output of countries. This was discussed in some detail in Chapter 5. Of the indicators listed there, we shall use here the number of scientific authors, that is, the number of scientists who published in a given year. The advantages and drawbacks of this particular measure are outlined in Chapter 5.

Taking 1973 as a year for which statistics are readily available, one finds that some 42% of the world's science was produced in the United States. The large West European countries—the United Kingdom, France, West Germany, and Italy—together accounted for another 22%, Japan 5%, and the Soviet Union 8%. Thus these seven countries produced slightly more than three-quarters of the world's science. Another 6% came from seven small European countries (Norway, Sweden, Finland, Denmark, Holland, Belgium, and Switzerland), and another 4% from seven East European countries (Poland, Czechoslovakia, East Germany, Hungary, Romania, Yugoslavia, and Bulgaria). Thus the above-mentioned 21 countries (about 15% of the total number of countries) were responsible for over 85% of the world's scientific literature.

These figures must be handled with some care. For one thing, they are based on a literature search that covered English-language journals more adequately than others and, in general, journals in one of the large Western languages more adequately than those in non-Western or small languages. Thus there may be some distortion in favor of the large Western countries. Second, the figures since 1973 have changed somewhat. For example, the margin of United States' leadership has decreased considerably. Yet, the above statistics are at least an approximate indication of the relative roles the various parts of the world play in creating new science.

Within the United States, there is also a pecking order among the states. California is the leading scientific state, housing, in 1973, 14,000 of the 114,000 scientific authors in the United States. The second is New York with 13,500, followed by Massachusetts, Illinois, and Pennsylvania. These five states among them contained 41% of all scientific authors in the United States.

We see, therefore, that the distribution of scientific contributions around the world and even within a given country is highly skewed: In that sense science may be regarded as elitist. In fact, looking at such figures, some may wonder how a greater equality among the various countries could ever be achieved, even if we calculate the number of scientific authors on a per capita basis. After all, India with a population three and a half times larger than the United States has only one-sixteenth as many scientific authors as the United States.

A Glimpse of the Future

It is not likely that these ratios will change substantially in the next few years. But if we look at longer time periods in history, and if we speculate on the extrapolations of present *trends,* we can see how these ratios might change. It is not likely that an equal distribution of scientific productivity will ever be attained, but the developing countries, as a group, despite the difficulties outlined earlier in this chapter, may rise in scientific production relative to the group of countries at the top of the list. Furthermore, some of the leading countries may soon lose their leadership and be overtaken by countries now far down on the list.

It appears from human history that great civilizations and great countries rise and fall and are overtaken by others. One of the surest signs of decline in the past has been a spreading fear, a feeling of uncertainty, a loss of confidence, a craving for security instead of a bold exploration of the unknown and a courageous assumption of risks. Consequently, it is likely that many leading countries are about to enter the autumn of their existence. Countries in which job security is paramount at the cost of encouraging the talented and capable, where many see danger in further scientific research rather than an opportunity and a road for an even greater fulfillment of human aspirations, or where catatonic bureaucracies are allowed to govern the life of the population, are ripe for decline.

In the past, the new barbarians managed to become leaders by taking over the leading countries, assimilating from them what was useful and valuable, and yet not being infected with the bacteria of decay. Whether such a change of roles can and will take place in the future is an intriguing question. The problem today is, in this respect, that communications around the world are so good that the negative and psychotic as well as the positive and valuable can spread over the world with amazing speed. In such a climate it may be more difficult for the aspiring leaders of tomorrow to make a selected assimilation of what emerges from today's leading countries.

Yet the key to the future for the now trailing countries is exactly in this rapid yet selective transfer. The great achievements of the now leading countries must be acquired before they become too tired and too lethargic to participate in the transfer. At the same time the indigenous capability of the ascending countries must be established, coupled with a fresh, new, progressive spirit, which can help serve as a recipient and a critical selector of this acquired legacy as well as a basis for achieving further progress. The collective, cumulative, objective character of science is likely to fare well in this difficult transfer process, so one can be quite confident that, one hundred to two hundred years from now, when a new hierarchy of countries has developed, science will continue to form an important aspiration of the new leaders.

APPENDIX A
HOW TO CORRECT ONE-DIMENSIONAL THINKING

WHO IS THE WORLD'S greatest composer of music? What is the reason for inflation? Who is the most beautiful woman alive today? What motivates people to undertake scientific research?

Such questions are heard often, and attempts are even made to answer them. But are these reasonable questions? To be more precise, is their structure such that they can be reasonably answered?

These questions, in fact, are not asked reasonably, and even trying to answer them distorts our thinking in a way that can be detrimental to real problem-solving in all walks of life.

In particular, these questions are products of one-dimensional thinking in areas that are multidimensional; therefore we must specify additional conditions and assumptions (which may vary from person to person) before we can give unambiguous answers.

To begin with, let us discuss a different kind of question: What is the highest temperature ever recorded in Canada? To answer this question, one needs to collect all the temperature readings ever taken in Canada (which, at least in principle, is quite feasible) and then compare them to find the highest.

The point is that temperatures can be measured along one scale, along one line on which there are marks for the different degrees of Centigrade (or Fahrenheit, if preferred, it makes no difference). We can describe this situation by saying that temperature is a one-dimensional concept. Consequently we have no difficulty comparing the various temperature readings with each other and determining the highest. This determination unambiguously answers our question, which we shall call a one-dimensional question.

Questions in Several Dimensions

Now let us turn to a question of another type: Minneapolis is 250 miles north and 300 miles west of Chicago, while Nashville is 400 miles south and 80 miles east of Chicago. Which is farther away from Chicago?

Note that this is no longer a one-dimensional question, in the sense that we have given *two* properties for each city: "north-southness" and "east-westness." In fact, Minneapolis outdoes Nashville in east-westness from Chicago, but Nashville outdoes Minneapolis in north-southness. And yet we certainly know that the question of which city is farther from Chicago makes sense and admits to an unambiguous answer.

Let us look at how this question is answered. Ignoring the clever people who will use the theorem of Pythagoras to calculate the two distances algebraically, we shall arrive at the answer in a simpler and more picturesque way. On a piece of paper (Figure 1) we draw a point signifying Chicago, and then measure 300 units up and 250 units left. (It does not matter how much a unit is.) We mark the point we arrived at as Minneapolis. Then we measure 400 units down and 80 units to the right, and call the resulting point Nashville. Then we take a compass and draw a circle with its center at Chicago and with Nashville falling on its circumference. If Minneapolis is within this circle, Nashville is farther; if Minneapolis is outside, it is farther.

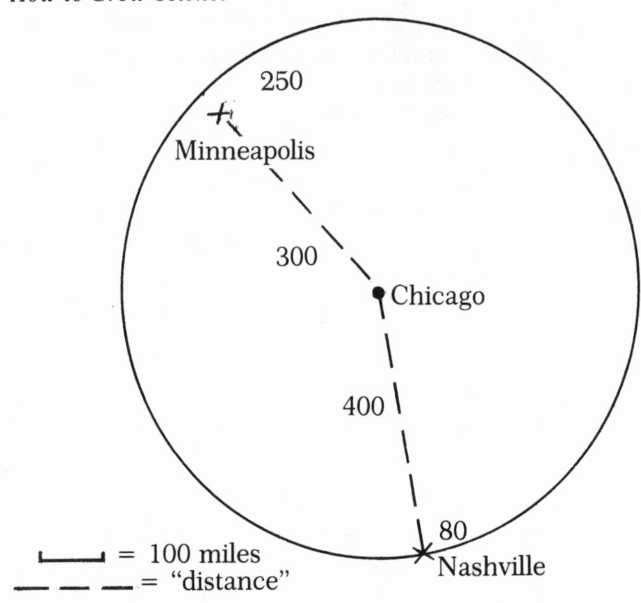

Figure 1

What permits us to combine in this fashion north-southness and east-westness? It is the fact that both these properties measure length and use the same units (miles, in our example). Thus, although we face a two-dimensional problem, the two dimensions are similar and commensurable, hence we can define a natural way of combining them. Thus, the two-dimensional nature of the situation still permits a one-dimensional answer.

Now let us progress to a more complicated question: Which is the best automobile made today? Take, as an example, two people, A and B, who are arguing about it. A is a racing driver, and B is a liberal-environmentalist-consumer. A likes to think in terms of speed, B in terms of durability. Yet they agree to consider both of these qualities in determining the answer to the question. So, having learned from the Minneapolis-Nashville example, they draw a graph on which the horizontal axis measures maximum speed, and the vertical axis the lifetime mileage. Both of these qualities can be given quantitatively in terms of numbers, so on that count there is no problem. Even so, however, A and B are unable to agree. Why? Using the method employed in the Minneapolis-Nashville problem, in comparing the two makes of Concorde and Sphinx, A and B draw the prescribed circles—yet there is no consensus. Concorde has a maximum speed of 225 miles per hour and lasts for 75,000 miles, while Sphinx has a maximum speed of 100 miles per hour and lasts for 150,000 miles. The two graphs, drawn by A and B respectively, are shown in Figure 2.

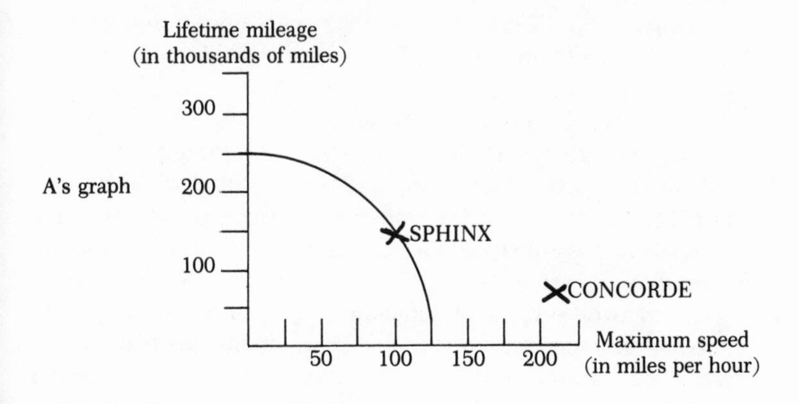

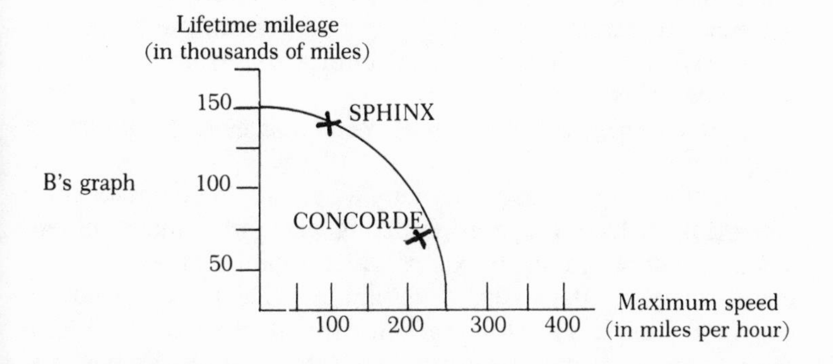

Figure 2

The difficulty is clear: Here we are also facing a two–dimensional problem and asking a one-dimensional question, but the two dimensions of the problem are no longer similar, no longer commensurate: They measure essentially different quantities and use different units in doing so. As a result, it is no longer possible to define an unambiguous, natural distance that combines the two dimensions, as was possible in the Minneapolis-Nashville problem.

One can, of course, define a distance (or magnitude), if one wants to, as A and B each have done. But this distance is no longer unambiguous, since it depends on a value system—on how many units of speed we equate with how many units of lifetime mileage. On A's graph, we see that 25 units of speed equal 50 units of lifetime mileage, while on B's graph 50 units of the former equal 25 units of the latter.

It follows from this that if someone is ignorant about the existence of many dimensions, and simply demands to have an unambiguous answer to the question of which car is best, he is asking an impossible question to which any set and unqualified answer will be wrong.

The idea of several dimensions is a mathematical concept embodied in a technical entity that mathematicians call a vector space. As is often the case, the technical subtleties and formalism of vector spaces need not be gone into by laymen in order to acquire the basic innovative idea for use in everyday thinking. Yet if the idea and simple properties of vector spaces were generally understood by every educated person, public discussions on almost any issue would immeasurably gain in clarity and practicality.

So what are these features of vector spaces that we need to keep in mind?

(1) When something has many essential characteristics, we can think of its evaluation as taking place in a many-dimensional vector space, each axis of which represents one of these characteristics. Naturally, it is difficult for us to visualize, concretely, such a vector space with more than three dimensions, since we are used to living in a three-dimensional geometric space (up-down, left-right, front-back). Yet we can try to think of an abstract vector space with any number of

independent axes, each representing a characteristic of interest.

(2) In order to be able to answer one-dimensional questions about objects in this many-dimensional vector space, we require, first of all, that the characteristic along each axis be measurable quantitatively—that is, each object with that property can be assigned a definite number on that particular axis.

(3) The condition given under (2), however, is not enough to enable us to answer a one-dimensional question, as our example of the two cars has shown. In addition, we must also have commensurability among the axes—that is, we must have a generally accepted way of making one unit on one axis equivalent to a certain fixed number of units on another axis.

In the absence of *either* of the two requirements discussed in (2) and (3), respectively, a one-dimensional question has no meaning and cannot be answered—that is, *any* unqualified answer given to it is wrong.

All this sounds rather evident or even trivial. Yet these simple observations are almost universally ignored. Virtually all real problems in life are multidimensional, yet almost all questions asked in public discussions, in politics, in cocktail party chats are one-dimensional. Correspondingly, most thinking is also one-dimensional. An example is the conspiracy theory of the world: "If only multinational companies were eliminated, economic disparities would disappear." "If only we could get rid of these environmentalists, the energy crisis would vanish." And so on.

If you are dubious, now that you are sensitized to spotting one-dimensional thinking, read your newspaper and listen to public debates with this in mind.

What are the dangers of such one-dimensional thinking? Perhaps above all, it radically oversimplifies situations, and hence not only searches for solutions in the wrong direction, but even looks with suspicion on any attempt to reveal the true multidimensional nature of problems. As an example, if someone claims that the motivation of undertaking scientific research is finding a cure to cancer *and* building more power weapons *and* enhancing national prestige *and* satisfying some basic drives in human beings to know and understand *and* modifying man's view of the world around him—and so on—he is likely to be viewed as a Machiavellian who is trying to perpetrate a fraud in several directions at once instead of as a realist who recognizes that the societal justification of sciences is highly multidimensional.

Another danger of one-dimensional thinking is that every difference automatically becomes a ranking as well. For example, differences in talents, capabilities, activities, or achievements between two races, or between two sexes, thus immediately become an explosive issue since such differences are inevitably connected with questions of inferiority or superiority. In contrast, when looked at in real, multidimensional space, differences do not unambiguously translate into a ranking, as we saw in the example of the two cars. Hence the introduction of any ranking is optional and relatively unimportant, and the results of such rankings depend on individual value judgments (in our terminology, on the relative sizes of the units along the various axes).

An Example

In addition to realizing, however, that a certain discussion is oversimplified and hence misleading, we may also want to use our new insight positively in order to arrive at an answer even in a multidimensional situation. Can we do this? Certainly not in all conceivable cases, because there may be elements in the situation that do not lend themselves to quantification at all, and hence points (2) and (3) cannot be applied to them. Yet there are fewer such elements than one might think at first. Here is an example.

If someone asks, "What is the monetary value of a human life?" we may tend to dismiss the question at first as callous and meaningless. "A human life is invaluable," we might answer. It turns out, however, that in practical situations we don't act according to this statement. We don't, for example, provide an infinite amount of medical care for members of society any more than we prevent people from driving cars because tens of thousands of people are killed each year in car accidents. In fact, by studying the extent to which we are willing to go to save a human life, we can, in accordance with point (3) in the proposed procedure, even place an approximate price tag on a human life. (It turns out to be somewhere in the neighborhood of $300,000, which is arrived at by ascertaining how much money society and individuals, on the average, spend on saving a human life.)

So, on the positive side, we may want to use our prescription as follows:

(A) Analyze the issue into its constituent elements, as explained in point (1) earlier.

(B) Try to quantify, even if only approximately, as many of these elements as possible. If we are willing to be satisfied, for some elements, with quite rough quantification, we should be able to quantify almost all elements of the problem, even if the problem is not a purely technological one.

(C) Find the relative scales for all the elements from which we managed to find a quantitative measure. We might find it difficult in some cases to do so, but for most elements this will be possible.

(D) We can construct a distance (as we did when comparing cities and cars) at least in terms of those elements of the problem that we managed to quantify and compare, hence we can then compare different statements offered by parties in the discussion. To be sure, this will not entirely resolve the problem, since nothing was done about the elements that could not be quantified, but at least we managed to get a partial answer and to narrow down the area in which true disagreement remains among the discussing parties.

Since all this sounds very abstract, here is a simpleminded example that may serve to illustrate the general idea:

Let us assume that we want to travel a distance of 500 miles, and we have two choices: driving our own car or taking a commercial plane. How can we compare these two possibilities?

First, we want to make a list of the relevant factors to take into account. They may be the cost of operating our car, the plane fare, safety, whether we feel comfortable in a car or a plane, the scenery we see on the way, etc.

Let us assume for the sake of simplicity that we decide that only three factors are relevant: the cost of the transportation itself, safety, and our comfort while on route.

Qualitatively, we know that the travel cost by car is lower, but that the plane is safer and our comfort is greater. Since there are advantages on both sides, a mere qualitative listing is not enough for a decision. Let us, therefore, try a quantitative evaluation.

If the plane fare is $80, and the cost of driving our car for 500 marginal (that is, additional) miles is 10¢ per mile, or $50 for the trip, going by car has a $30 advantage on that score. What about safety? If we use the aforementioned $300,000 per human life, and knowing that the fatality rate for private cars is 3 per 100 million passenger miles while it is 0.1 per 100 million passenger miles for commercial flights, we can easily conclude that the cost of our safety risk for that trip is 15¢ on the plane and $4.50 by car. Thus, so far, on account of the first two dimensions, driving by car is still ahead by about $25.

The last dimension is that of comfort, and here we have to make a value judgment about how much comfort is equivalent to $25. This is a personal judgment which will then decide whether to take the plane or the car.

Someone may object that since at the end a personal value judgment decided the matter anyway, there was no point in analyzing the other dimension quantitatively. That is not quite so, however. The quantitative assessment of the other two dimensions accomplished two objectives. First, it showed that an absolute objection to going by car on the basis of safety had no factual ground, since the danger of going by car could be measured and shown to be relatively negligible. In practice, even in the face of this evidence, we may want to make the choice of going by plane because it is safer, but if we do so, we know where our decision came from—from an overwhelming concern with safety—and hence a rational argument about the relative merits of the two alternatives will be pointless.

Second, since we managed to quantify all other dimensions, the assessment now placed a definite financial tag on our extra comfort and hence made it easier for us to decide whether we want to spend that amount of money for the extra comfort.

As mentioned, this example is too simple, since most people would not bother to think about such a relatively trivial decision in such elaborate terms. If, however, we were faced with a much larger problem, with many dimensions, with many people participating in the decision-making, and with millions of dollars potentially involved, such an analysis could be much more helpful.

In summary, we can say then that the realization of the multidimensional nature of problems and the ensuing appropriate analysis of such problems may in some cases bring about a definite, unambiguous solution, and even in other cases will help us to separate those dimensions of a problem that depend on value judgments. As a result, we can narrow down further discussion of a problem to a dialogue between two value systems along only a few of the many dimensions of the problem. This will simplify, or at least clarify, the controversy that surrounds the problem.

APPENDIX B
*EVOLVING THE ORGANISM OF SCIENCE: APPLYING A FABLE**

ONCE UPON A TIME there was a man who lived in a little house. He was a poor man and had to work hard to provide the bare necessities for survival. His hut was on a flat piece of land, surrounded by weeds and scrubby brush, just as it was when he came to settle there.

When, at the end of a long day's work, he sat down to rest after his evening meal and looked out the window at the jumble of greenery surrounding him, he felt dissatisfied. It surprised him that this should be so, even with a full belly and a chance to find some leisure. But then he realized that he was unhappy because he remembered a trip to the mountains and recalled the miraculous feeling of beholding tall trees overhead, the awe inspired by their reaching to the sky, the pleasure of listening to the rustling of leaves. That wonderful experience was in sharp contrast with the stark reality of the ugly weeds around his house.

He decided that he must have some trees. It was not easy to

*This article appeared in *Leonardo, 11,* 214 (1978). Reprinted with permission.

turn this resolution into reality. He spent days and weeks observing the trees in the forest and trying to transplant them into his yard. And finally he succeeded. Trees began to grow where only weeds had been, and he felt happier, even though he was still struggling hard just to survive.

Years passed and the trees matured, and one year, to his amazement and surprise, they began to yield fruit. The fruit was large and juicy and most desirable to look at and to eat. Year after year the amount of fruit multiplied, and he became wealthy and well nourished, eating some of the fruit and selling the rest. As he grew older, he enjoyed an easier, more pleasant life.

About that time a young man moved to the area and built a house next to his. He was a poor fellow and, like his neighbor in his younger days, worked hard to survive. As he sat, after a long day's work, on the porch of his house, he kept gazing at the wonderful fruit growing across the fence in his neighbor's yard. He envied the neighbor for being so rich. From time to time, he also felt how nice it would be to be shaded by those beautiful green leaves and waving branches, but he quickly dismissed this feeling as not appropriate for a poor man.

Finally, he resolved that he must also have some fruit like that and he must have it quickly. He reasoned that, since he was poor and getting older, he could not wait much longer. He decided, therefore, that he would produce the fruit and the fruit only. His reasoning was simple. First, he decided that most of the roots, branches, and leaves on the trees were useless, so he could reduce their number. Let the rich bother with waste. Second, he came to the conclusion that with the neighbor's trees already there, he could avoid the arduous task of transplanting trees from the forest.

He asked his neighbor for some twigs from the fruit trees and stuck them into the ground in his yard. But no fruit was produced; in fact, the twigs wilted and died. He tried again; only this time he obtained small saplings from his neighbor. They began to grow, and so he proceeded to tend them in his own way. He trimmed the roots from time to time, so they should not reach too far into the ground and use up too much nourishment and water. As small branches and some leaves began to grow, he cut off all branches but one (the one that looked similar to the branch carrying fruit on his neighbor's tree), and on that branch he carefully trimmed the leaves so that nothing useless would be produced by his trees.

But the result was a failure. His trees did grow, but they had weird shapes and yielded no fruit, in spite of his spending much time tending them. Year after year, whenever he wanted to eat fruit, he had to squeeze some money out of his meager income to buy some from his rich neighbor.

This failure was a heavy burden on the young man's mind. It was not only a matter of having to pay for the fruit he wanted. He increasingly realized that the burden arose mainly from a feeling of inferiority, from a sense of competitive impotence relative to his neighbor.

He finally became weary of all this and decided to have a long talk with his rich neighbor to find out the secrets of growing trees that richly bear fruit. His neighbor, however, did not tell him any deep secrets. He said nothing about magic nutrients, clever tricks, and sophisticated procedures that assure fruit on the trees. All he told him was his own life story, his desires and motivations in growing trees in the first place, and his surprise at eventually finding the fruit on the trees. Then, in a reflective mood, he added: "I think my real fortune was in not having had a neighbor whose tempting fruit trees I could behold. If I had had one, I should easily have been led astray, just as you were by my trees. When one wants to reproduce the results of others, one is often blinded by them and fails to see the complexity of the essence that brought about those results. Not having such results to replicate, I was lucky enough to have been carried away with an aspiration to catch the essence and thus was rewarded also with the results."

Applying the Fable

What is commonly called the modern scientific revolution began in Western Europe about four hundred years ago, when some pioneering minds became dissatisfied with the static, fatalistic outlook of the Middle Ages. During the Middle Ages what was believed to be true about the world was transmitted by a few specialists who consulted texts by accepted authorities, including the belief that the world was governed by the will of the Christian God and that humans were only His passive pawns. Through the efforts of these pioneers, strong aspirations arose to reshape the prevailing mental picture of the world on the basis of direct observation and experimentation—in other words, to forge a new world view.

The material improvement of human life was not the motivation underlying the modern scientific revolution. Little did those pioneers realize what a stupendous impact the new way of learning to understand the world would have in this respect. Actually, this impact was not felt until about three hundred years after Copernicus (1473–1543), and the reason is easy to understand.

In the early days of modern science, investigation was focused on natural phenomena that are directly available to the senses for observation. Thus the branches of physics dealing with mechanics (the effect of forces on solids, liquids, and gases), light, and sound were first pursued. These branches have a direct bearing on many areas of technology, but tools, pumps, steam engines, and mechanical weapons could be and were developed empirically—that is, by trial-and-error methods rather than by those of the new science. Since the invention of simple hand tools in the distant past, and until about one hundred years ago, empirical methods were used in technology. Even today, some technological innovations are made by clever, resourceful people without a scientific background, provided the innovations involve phenomena directly available to the senses.

Until about one hundred years ago, the motivations of scientists were mainly of a philosophical, intellectual, and cultural character, and these motivations, as an inquisitive methodology, are considered to underlie basic or fundamental research. Thus science in Western civilization had had two to three centuries to implant in societies a new world view, in the sense discussed above, in terms of these motivations. Since then, science has acquired a broader scope. Scientists began to give attention to aspects of nature that are not directly accessible to the senses (electric phenomena that cannot be either seen or heard; atomic physics and then particle physics, chemical processes, and microbiology, which involve phenomena that cannot be perceived directly). Knowledge of these phenomena cannot be obtained by solely empirical methods, and inventors without a scientific background are not likely to contribute to a technological exploitation of these domains.

In principle, if an imaginative person without a scientific background was provided with a box containing pieces of copper, germanium, plastics, steel, wood, and glass, he might, by trial and error, eventually invent a transistor radio, but it would require an effort of many millions of years! By 1900 it became apparent that further technology will depend on persons who grasp the understanding of nature provided by science, and this has led to a new motivation for doing science—namely, applied scientific research. By then, however, the subtleties of how to pursue science were learned and fairly widely absorbed as a general kind of human activity. This new motivation could be incorporated into the already existing structure of basic science, and science policies of governments and individuals now include technological motivation as *one* of the justifications for doing science.

Recently, however, some aspects of Western civilization have tended to spread throughout the world. Sporadic signs of this trend began about a hundred years ago, and they became clearly evident during the past fifty to seventy-five years. A notable example is Japan, where the trend started a bit more than one hundred years ago, and where a number of favorable circumstances supported it. However, in other so-called developing countries, with or without a Western colonial experience, one can speak about a substantial influence of and the beginning of the growth of modern science and advanced technology only during a shorter period of time. For the purposes of discussion, the important factor to note is that this influence began when science and advanced technology already were visibly related to each other—that is, when the technological motivation for pursuing science was broadly accepted.

The first influence on developing countries came from advanced technology in the form of automobiles, radios, and other kinds of consumer products. Manufacturing methods were transferred so that the production of such goods, with relatively low labor costs, would improve the foreign trade balance of a country. Thus the impression in these countries was strengthened that through the mysterious thing called science and technology they could obtain political power and wealth commensurate with that of advanced-technology countries.

Science as an intellectual and cultural force also began to seep into these countries, but very slowly. Up to now it has affected only a small number of those who were educated in science and advanced technology abroad. They had the talent, inclination, and nonconformist attitude that permitted them to absorb the scientific spirit and make it an integral part of their lives. In most developing countries, however, these are not the persons who initiate governmental policies for science and technology, just as in the West the early pioneers of science did not play a large role in the social and political development of their societies.

Those who do make the decisions on science and technology were exposed to an education that shielded them from "unnecessary and useless" ideas and views and that did not include an introduction to the historical, philosophical, and sociological aspects of science. Their attention was directed to the fruit in the neighbor's garden of our introductory story. They are, thus, obsessed with the view that, since they are responsible for improving the condition of the people in their poverty-stricken countries, they should not indulge in anything in science that does not promise immediate material results. They formulate science policies that trim the roots, cut the spurious branches, and eliminate the useless leaves in the hope that fruit will be produced more quickly and cheaply. Such an approach can lead only to failure.

This myopic approach is not a consequence of the poverty that prevails in most developing countries. Several of them have rich traditions of making "useless" products—for example, temple wall paintings in Thailand, wood carvings in Nigeria, masks in Zaire, and batiks in Indonesia. The contrast between such manifestations and the blindness to the philosophical and esthetic aspects of science is striking but not surprising, since art for magical, religious, and decorative purposes has been practiced in these societies for centuries, but science was introduced only a few decades ago.

We see blindness and incomprehension not only regarding the essence of science but often seemingly also regarding its results. The functional evaluation of the results of scientific research is almost completely missing in the developing countries. What their governments require, instead, are vacuous formalistic monthly progress reports, which can be conveniently placed in filing cabinets. In view of their expressed desire to produce fruit like the neighbor's, such indifference to the evaluation of the results is at first astonishing.

It is not so astonishing, however, since, as in the case of the young man of our story, the desire in the developing countries for growing fruit like the neighbor's is not mainly a result of the attempt to have fruit to eat but a craving not to be inferior in fruit-growing to the neighbor. The main motivation for development is not to acquire material wealth but to eliminate a sense of inferiority in a world containing advanced-technology countries. As, over the years, I visit scientific and technological institutions in many developing countries and am given guided tours complete with commentary, it is virtually impossible for me not to conclude that the game in these countries is to exhibit the existence of science and technology to demonstrate that they are equal partners in the world community.

Such motivation is likely to play a decisive role in determining the direction of development in these countries. It should not be allowed to become the dominant one, however. If it does, fruit that looks like the neighbor's may result, but it will be sour, wormy, and inedible.

For the scientific revolution to pervade a country what is needed is long-term, broadly based education for future generations, the energetic support of the few individuals who have already been initiated into the scientific spirit, and a steady interaction by these countries with persons from advanced-technology societies where the scientific revolution has, at least to a considerable extent, been absorbed in the general culture. Stress should be placed on the benefits to be gained from science in molding a new outlook on the world, rather than only on its contribution to making physical survival easier. In this respect, one may encounter resistance even from some Western development experts (especially political economists) who exhibit an astonishing ignorance about what it takes to introduce productive science in developing countries.

Intellectual and philosophical transformations of societies, drastic change in a people's outlook on the world, and new cultural trends cannot be produced quickly. There is no known way to grow the tree of science so that it produces fruit a week after planting, but unfortunately there are many ways of tending the tree so that no fruit will ever be collected. If the unavoidable features of introducing science and advanced technology could be convincingly communicated to eager people in developing countries, these people would avoid many future disappointments and gain material as well as other benefits.

APPENDIX C
*TWO WAYS OF MOUNTAINEERING**

SUPPOSE YOU LIVE in a small town near a mountain range and want to become proficient in mountaineering. One way is as follows: To start, you put on some old tennis shoes and walk up the nearest hill. Perhaps you puff by the time you make it, but the next time, and the time after, it is easier and you get up faster.

Next, you get a small knapsack and wear a pair of stronger shoes and attempt the closest mountain of moderate height. Again, after a few tries, this goes well, and you are then ready to get hold of a pup tent and a sleeping bag and venture for an overnighter. In this you may want to get advice from some mountaineer friends who have done it before. Soon you are quite skillful and confident and decide to embark on some rope climbing and glacier crossing, in the company of more experienced friends who are willing to lend you ropes and crampons.

Eventually you graduate to even more challenging technical achievements. All through this process, when anybody asks, you glowingly relate how high you have gotten, how many miles you have covered, how demanding a rock wall you have conquered.

*Reprinted with permission from *Physics Bulletin*, 29, 249 (1978). Copyright © 1978 The Institute of Physics.

Never during this process do you lose sight of the main purpose, that of reaching the mountain tops and becoming an accomplished mountaineer.

There is, however, another way. You start by buying handbooks for mountaineers describing in great detail the techniques of organizing expeditions. You follow this by contacting and joining the Universal Expedition, Safari and Climbing Organization (UNESCO), receiving its magazine and obtaining its equipment guides.

You then begin investing in arctic tents, Himalayan packs, ice axes, space age freeze-dried food, and expeditionary climbing boots. You also draw up elaborate plans for your future expeditions, including coordinating managers, guides, supply organizations, and time schedules, and even establish an office for such activities with the prominent banner of Centre for Reaching Alpine Peaks (CRAP). During all this you find no time to visit even the nearest hilltop, but when queried you proudly boast about your shiny new equipment and busy Centre, give stirring speeches about the importance of mountaineering, and sometimes even parade up and down the street in your new climbing outfit.

Ridiculous? Obviously, in this context—but apparently not ridiculous enough when these two approaches are projected into a different, perhaps less familiar, area.

Suppose you want to build up and utilize science and technology in a country where there is practically none. Again there are two ways.

You can start by providing your few scientists and engineers with an environment where they can exert their talents and energies. After a while, this manpower expands, attains variety, and solidifies its direction and orientation, and it has a greater impact on the rest of the country.

As already existing equipment becomes clearly inadequate for the work in progress, new acquisitions are made. Eventually the community becomes large enough to need some firmer administrative structure, which you provide on a minimal level as the necessity for it arises.

You strengthen your community by assisting interaction with colleagues locally and abroad. All this time, you assess this development by marking the rising output of your manpower in terms of research papers, reports, patents, contributions to production methods, and the like.

There is, however, another way. You start with a Grand Plan for Science and Technology for Development, involving organizational charts, targets, and chains of command. You erect splendid buildings and acquire shiny new equipment, and in these national centers of research you place administrators.

You join international organizations and send your bureaucrats to international meetings on science policy where they learn about the use of the Delphi method in policy formation and agree to create yet another international organization.

You distribute research funds according to whether the title of a project appears to harmonize with the national plan, and whenever someone asks you about the development of science and technology in the country you are careful to cite only input quantities: the number of buildings and administrative bodies, the amount of equipment, and perhaps the number of people with degrees in science and technology (regardless of quality).

Whether, amid all these formalistic manipulations, costly and time consuming as they are, any science or technology is produced or utilized is not asked and not assessed.

At present, the second approach is all too common; hence the goal of the development of science and technology and their utilization suffer in many countries. When will the worldwide scientific and technological community take the initiative to recast the perspectives and reassess the approaches?

APPENDIX D
HOW CAN TWO SCIENTISTS DISAGREE?

IN THE MINDS of many laymen a scientific judgment is objective, unambiguous, and hence authoritative. It is therefore confusing to them when, as often happens in the public arena, two scientists make diametrically opposite public statements about a current issue. To many, in this changeable and unreliable world, a scientific utterance represents the only fixed, solid, permanent point of reference—and then this illusion of security is shattered by two scientists' blatantly disagreeing with each other.

Such a disagreement can come about very easily, indeed, and in fact can have a number of different causes, singly or in combination.

The *first* possibility is that the issue in question has nothing whatever to do with science. In that case, of course, the word of a scientist is, at best, as good as that of the owner of the nearest grocery store, and there is no reason in the world why two scientists should not have two different opinions, one or both of which are fallacious or born in ignorance, anger, or fear.

As an example, recently, shortly after taking office, a state governor decided to appoint new members to the energy siting board which determines the location of nuclear power plants. In reaction to this move, brandishing their scientific credentials, three distinguished scientists made a public statement condemning the governor's move. The issue in question, of course, had nothing to do with science, and thus the "scientific" protest was made under completely false pretenses.

A *second* possibility is that, although an issue is a scientific one or has significant scientific components, one or perhaps both of the scientists who disagree don't know much about that particular scientific area. After all, science is huge, and nowadays one person can aspire to have thorough knowledge of only a tiny corner of science. A scientist may also have some understanding of other areas of science, but not sufficiently to be able to make an authoritative statement.

There are many examples of such situations, including instances on both sides of the controversy about the safety of nuclear fission reactors. That subject involves many quantitative details pertaining to biology, physics, and chemistry, as well as engineering, economics, military science, and other subjects, and it takes years of full-time devotion to master the subject sufficiently to be able to make reliable statements. Naturally, many scientists make statements even in the absence of such a thorough background, but if they do, these statements should not be taken at face value and certainly should not be expected to be free of ambiguities and contradictions.

A *third* reason for the disagreement may be that although the issue in question is a scientific one, the scientific solution of it is not yet known. There are many scientific problems that we can formulate on which research has not been performed and hence for which we don't know the answers. To take a somewhat ludicrous illustration, we do not know whether or not eating ten hard-boiled eggs on top of Mount Everest causes cancer. We have no reason to believe that the answer will be in the affirmative—that is, the plausible mechanisms we can think of would suggest that the answer is no. This, however, may not placate either the layman or the scientist if he fears the unknown. Since cancer is hardly understood at all, fearmongers can simply join their ancestors in the Middle Ages to whom the world was full of evil spirits. One often hears calls for the "absolute safety" of a product or a technological gadget before permitting its use, but such an absolute assurance can never be given, either in principle or in practice. There may be some science that we do not yet know that is pertinent to the issue.

On some problems, however, practically no research has been performed, so scientists (as well as nonscientists) are completely at liberty to speculate on the outcome of the research.

To proceed now to a *fourth* situation in which two scientists can be reasonably expected to disagree, consider an issue (typical of many or even of most that arise in practice) that is a combination of scientific and nonscientific factors, and where the scientific part of the issue is reasonably under control but the nonscientific part is characterized by lack of knowledge and information. For example, whether we should project nuclear or coal-burning power plants for a time fifteen years from now depends not only on scientific but also on economic factors, and since economics is not well enough developed to be able to predict, with any certainty, the relative costs, availabilities, distribution patterns, and social acceptabilities of uranium versus coal fifteen years from now, considerable disagreement is possible between two scientists, although not on the scientific but on the economic components of the issue. In such a situation, if the scientists were honest they would preface their public remarks with a brief explanation about the lack of scientific disagreement between them and about the existence of a difference in views only because of the vagaries of economic projections. In actual life, very few scientists are careful and modest enough to offer such explanations, and so to the public it appears again that the two scientists differ in their views as scientists, which is not the case.

A somewhat related *fifth* possibility is that two scientists disagree on the nonscientific components of an issue not because knowledge and information are lacking but because the scientists are ignorant about these nonscientific components. Often, successful scientists fancy themselves also experts in nonscientific fields, and hence do not hesitate to make authoritative-sounding statements on issues that are a combination of scientific and nonscientific components and in which there is little disagreement on the scientific aspects. As mentioned above, scientists in this case also are less than careful in making a distinction between scientific and nonscientific components of a problem, and hence the public may be confounded by the scientists' disagreement.

A somewhat related situation arises, as a *sixth* possibility, when the nonscientific component of an issue does not lack factual support or a knowledge base but instead is highly value-dependent. For example, in the public controversy about the energy crisis, an important part is played by various conceptions of what constitutes a good, rich, rewarding, and exciting life. To some, the idyllic image of people living in a world uncontaminated by modern technology is attractive, while others find their aspirations in supersonic flight, space exploration, and artificial intelligence. For this reason, public discussions about the quality of life are almost always worthless, since this phrase is dear to everyone and has a different meaning for each of us. Scientists, therefore, may very well debate issues with some scientific components, the other factors of which depend on personal values, and since in this situation, again, scientists will almost never admit that their disagreement is due to a difference in personal values and not in the assessment of the scientific components, the public is again mystified.

A *seventh* scenario in which two scientists may disagree involves a purely scientific issue in which the scientific ingredients are not unknown, but about which the pertinent scientific knowledge is qualitative and not quantitative. An example of this is the controvery over whether or not supersonic planes will destroy the earth's ozone layer. In this case it is possible to list, without serious disagreement by any scientist, chemical reactions which feed on substances produced during a supersonic flight and which deplete the ozone. What is quite unknown, however, is the quantitative extent of this depletion. This lack of information is crucial for the policy aspects of the issue, because if the rate of depletion is such that we may lose 1% of the ozone in ten thousand million years, we do not have to concern ourselves, while, if the depletion is 1% in one hundred years, there may be reason to pay attention to the effect. In the absence of reliable quantitative information, the public utterances of scientists may well be determined by nonscientific, psychological, and personal factors, and so disagreement is very possible.

An *eighth* possibility may be illustrated by a situation in which even quantitative scientific information is available, but not quite in the range of the parameters that occurs in the issues at hand. In this case some extrapolations must be made. These extrapolations can be done in a number of ways, and which of these is plausible to someone may very well depend on nonscientific factors. For example we know little about the long-term effects of extremely low-level nuclear radiation, but we do have by now reasonably reliable data on the long-term effects of higher-level radiation. We therefore must extrapolate our data down to very low levels. The extrapolation is done partly mathematically and partly in view of what we know about the mechanisms of interaction of radiation with biological tissue. Thus, in this case, most scientists competent in this field agree on how to do the extrapolation, but there are some who, most likely for nonscientific reasons, disagree; and since actual data on such low-level radiation are not yet available, it is impossible to prove that they are wrong. Certainty in any case is illusory. It may be possible to say that a maverick's being correct has a chance less than one part in a trillion, but that will not silence him, and hence, to the public, the disagreement will stand.

Finally, a *ninth* reason why two scientists may publicly disagree on an issue is connected with all information and knowledge in science having a certain range of uncertainty. For example, if we measure the length of a table and find it to be, say, 1 meter 67 cm., the accuracy of this statement is never infinite. The measuring tape, after all, may not be absolutely correct, it may have stretched a bit, the marking on it may not be fine enough so that we can read the length to the nearest millimeter, etc. Thus in any scientific measurement the result must be given with a definite uncertainty, or error estimate, attached to it. In this example, we may say, for example, that the length of the table is 1 meter 67 cm. 4 mm. (± 1 mm.), which means that we are fairly sure only that the result is somewhere between 1 meter 67 cm. 3 mm. and 1 meter 67 cm. 5 mm.

If many such measurements (as well as approximate theoretical calculations) are put together into the scientific evaluation of a problem, the many small uncertainties combine and the final result will also have a certain error estimate.

All this means that even if science understands a problem both qualitatively and quantitatively, the answer it can give is quantitatively reliable only within a certain band of uncertainty.

In discussing a particular issue, therefore, two scientists may choose, for personal, nonscientific reasons, the opposite limits of the band of uncertainty and hence possibly come to different conclusions concerning public policy.

The foregoing nine reasons or possibilities for scientists to disagree on an issue cover most practical situations. In any given case, however, several of these reasons may be operating in conjunction, so the analysis may be more involved.

Besides possibly having some entertainment value, why is it useful to prepare such a list of possibilities?

First, it is useful for demonstrating that the objective, universal, and unambiguous character of science is not necessarily in contradiction with the possibility of a disagreement between two scientists. But beyond that, such a listing may also help the layman to clear up confusion in particular instances. The layman must ask the scientists questions that help to make a distinction among these possibilities, so that the true reason for the disagreement will surface. The scientists themselves cannot be relied on to supply this information without being pressed. It is only human that the great temptation of airing personal views and prejudices through a platform erected for objective, unambiguous scientific evaluations cannot be resisted. Only intelligent outside probing can therefore reveal what is behind the disagreement. Such a revelation can then help the layman to form a much better considered opinion on the issue, and may also point in the direction in which possibly more research, more debate, or more persuasion is necessary.

BIBLIOGRAPHY

AS WITH MOST BIBLIOGRAPHIES, the problem here is not what to include but what to leave out. Considering the likely readership of this book, I have followed some rules which, though perhaps somewhat arbitrary, appear to be functional. Although there are many interesting journal articles related to the topics discussed here, I have listed only books—as well as articles in a recent "encyclopedia" of social studies of science, because they themselves contain extensive bibliographies. Most are systematic studies rather than popular books, since the latter are more likely to be known to the average reader; yet I hope that the typical reader of this book will find most of them readable and interesting. Some are probably somewhat more difficult than this book, but there should be many that will cause no difficulty. The inclusion of an item, however, does not necessarily signify my agreement with or approval of its content.

Barber, Bernard. *Science and the Social Order*. New York: Free Press, 1952. An early classic in the sociology of science.

Ben-David, Joseph. *The Scientist's Role in Society*. Englewood, N.J.: Prentice-Hall, 1971. Thoughts on the relationship between history and science.

Bernal, J. D. *The Social Functions of Science*. Cambridge, Mass.: MIT Press, 1967. A reissue of one of the earliest major works on the societal context of science.

Blume, Stuart. *Towards a Political Sociology of Science*. New York: Free Press, 1974. An illustration of one of the contemporary trends in the sociology of science.

Brooks, Harvey. *The Government of Science*. Cambridge, Mass.: MIT Press, 1968. A sample of the writings of a prominent figure in science policy studies.

Bunge, Mario. *The Myth of Simplicity*. Englewood Cliffs, N.J.: Prentice-Hall, 1963. An analysis of simplicity as an element in the scientific method which attempts to debunk what it considers a myth, namely that simplicity is a guideline for scientists to select theories.

Bush, Vannevar. *Science, the Endless Frontier: A Report to the President on a Program for Postwar Scientific Research, 1945*. Reissued by the National Science Foundation, Washington, D.C., 1960. An essay that was very influential in shaping U.S. science policy after World War II.

Cantore, Enrico. *Scientific Man*. Forest Grove, N.J.: International Scholarly Book Services, 1977. An eloquent discussion of science from a humanistic point of view.

Cole, Jonathan, and Cole, Steven. *Social Stratification in Science*. Chicago: University of Chicago Press, 1973. A well-known book within the "modern" sociology of science.

Conant, J. B. *Science and Common Sense*. New Haven: Yale University Press, 1951. A popular book by a statesman of science.

Crane, Diane. *Invisible Colleges*. Chicago: University of Chicago Press, 1972. A significant contribution to the study of the informal structure of the scientific community.

Dyson, Freeman. *Disturbing the Universe*. New York: Harper & Row, 1979. An elegant mixture of autobiography and commentary by one of the unorthodox scientific minds of our time, who at the same time has the breadth of background to see things in a larger perspective.

Fermi, Laura. *Atoms in the Family*. Chicago: University of Chicago Press, 1961. A personal account by the wife of one of the great physicists of the 20th century.

Friedlander, Michael W. *The Conduct of Science*. Englewood Cliffs, N.J.: Prentice-Hall, 1972. A popular book with an objective somewhat similar to that of this book.

Garfield, Eugene. *Essays of an Information Scientist*. Philadelphia: ISI Press, 1977. A kaleidoscopic collection of brief comments by the central figure in the Institute for Scientific Information.

Gaston, Jerry. *Originality and Competition in Science*. Chicago: University of Chicago Press, 1973. An interesting case study of a scientific discipline.

Goudsmit, S. A. *Alsos: The Failure of German Science*. New York: Henry Schuman, 1947. An exciting account of the "science race" during World War II.

Graham, Loren. *The Soviet Academy of Sciences and the Communist Party*. Princeton: Princeton University Press, 1967. An account by one of the leading scholars interested in Soviet science.

Greenberg, Daniel. *The Politics of Pure Science*. New York: New American Library, 1967. A somewhat journalistic but forceful exposition of science and politics.

Haberer, Joseph. *Politics and the Community of Science*. New York: Van Nostrand-Reinhold, 1969. Another view on science and politics.

Hadamard, J. S. *An Essay on the Psychology of Invention in the Mathematical Field*. Princeton: Princeton University Press, 1945. A fascinating account by a great French mathematician of how scientific ideas are born.

Hagstrom, Warren. *The Scientific Community*. New York: Basic Books, 1965. An often-quoted discussion of the sociologist's view of the scientific community.

Hanson, Norwood R. *Patterns of Discovery*. Cambridge, Mass.: Harvard University Press, 1965. A stimulating contribution by an unorthodox mind.

Heisenberg, Werner. *Physics and Beyond*. New York: Harper & Row, 1971. A most educational and often philosophical personal account of the interactions among some of the great figures of 20th-century physics.

Holton, Gerald. *Thematic Origins of Scientific Thought*. Cambridge, Mass.: Harvard University Press, 1973. An eloquent analysis of the development of science by a physicist-historian-methodologist in terms of individual ways of looking at phenomena (themata) which strongly influence the positions scientists take in controversies.

Kaplan, N. *Science and Society*. Chicago: Rand McNally, 1965. A well-known book on the societal environment of science.

Kuhn, Thomas. *The Structure of Scientific Revolutions*. 2d ed. Chicago: University of Chicago Press, 1970. The book that gave respectability to the old idea of the scientific breakthrough.

Lakoff, Sanford. *Knowledge and Power*. New York: Free Press, 1966. A good account of the effect science has on many aspects of human life.

Laudon, Lawrence. *Progress and Its Problems: Toward a Theory of Scientific Growth*. Berkeley, Calif.: University of California Press, 1977. A point of view on the growth of science which often runs counter to how scientists themselves see it.

McCain, G., and Segal, E. M. *The Game of Science*. Belmont, Calif.: Brooks/Cole, 1969. The aim of this book is somewhat similar to this one.

Meadows, A. J. *Communications in Science*. London: Butterworth, 1974. A fact-filled summary of communication patterns in science.

Menard, Harry. *Science: Growth and Change*. Cambridge, Mass.: Harvard University Press, 1971. A good example of scientometrics. A quantitative discussion of the dynamics of science.

Merton, Robert. *The Sociology of Science*. Chicago: University of Chicago Press, 1973. A collection of contributions by this prominent sociologist.

Mitroff, Ian. *The Subjective Side of Science*. Amsterdam: Elsevier, 1974. A sample of writing by a man who likes to emphasize the nonobjectivity of science.

Moravcsik, Michael J. *Science Development: The Building of Science in Less Developed Countries*. Bloomington, Ind.: Pasitam, Indiana University, 1976. A survey of the rationale of and problems in the development of science in the Third World.

Mulkay, Michael J. *The Social Process of Innovation: A Study in the Sociology of Science*. London: Macmillan, 1972. An example of writing by a prolific British scholar.

National Science Board. *Science Indicators*. Washington, D.C.: U.S. Government Printing Office, 1977. An example of the quantitative data describing science and technology.

Nelkin, Dorothy. *Nuclear Power and Its Critics: The Cayuga Lake Controversy*. Ithaca, N.Y.: Cornell University Press, 1971. A sample of the writings of a prominent contributor of case studies on the interaction of science and society.

Polanyi, Michael. *Personal Knowledge*. London: Routledge & Kegan Paul, 1958. A central work by a noted interpreter of the nature of science.

Popper, Karl. *Conjectures and Refutation*. London: Routledge & Kegan Paul, 1963. A book by a prominent figure in scientific methodology.

Price, Derek de Solla. *Little Science, Big Science*. New York: Columbia University Press, 1963. A popular book describing how to measure science.

Price, Don K. *The Scientific Estate*. Cambridge, Mass: The Belknap Press of Harvard University, 1965. A well-known book on science and society.

Primack, J., and Von Hippel, F. *Advice and Dissent: Scientists in the Political Arena*. New York: Basic Books, 1974. A personal view of two physicists who turned to politics.

Ravetz, Jerome R. *Scientific Knowledge and Its Social Problems*. Oxford: Clarendon Press, 1971. A contribution by a noted British scholar on science.

Salomon, Jean-Jacques. *Science and Politics*. London: Macmillan, 1973. A perceptive book by a well-known French figure in science policy.

Sheinin, Y. *Science Policy: Problems and Trends*. Moscow: Progress Publishers, 1978. An example of the formalistic science policy studies fashionable in centrally planned economies.

Shils, Edward (ed.). *Criteria for Scientific Development: Public Policy and National Goals*. Cambridge, Mass.: MIT Press, 1966. An assortment of practical topics in science policy.

Skolnikoff, Eugene B. *Science, Technology and American Foreign Policy*. Cambridge, Mass.: MIT Press, 1967. A well-known book by a noted political scientist.

Snow, C. P. *The Two Cultures and A Second Look.* Cambridge: Cambridge University Press, 1969. A much-discussed essay on the cultural gap between scientists and nonscientists.

Spiegel-Rösing, Ina, and Price, Derek de Solla (eds.). *Science, Technology, and Society*. London and Beverly Hills, Calif.: Sage Publications, 1977. Contains the following recent reviews:

Bohme, Gernot. "Models for the Development of Science."

Fisch, R. "Psychology of Science."

Freeman, Christopher. "Economics of Research and Development."

Lakoff, Sanford. "Scientists, Technologists, and Political Power."

MacLeod, Roy. "Changing Perspectives in the Social History of Science."

Mulkay, Michael J. "Sociology of the Scientific Community."

Nelkin, Dorothy. "Technology and Public Policy."

Ravetz, Jerome R. "Criticism of Science."

Salomon, Jean-Jacques. "Science Policy Studies and the e-velopment of Science Policy."

Sapolsky, Harvey M. "Science, Technology, and Military Policy."

Schroeder-Gudehus, Brigitte. "Science, Technology, and Foreign Policy."

Skolnikoff, Eugene B. "Science, Technology, and the International System."

Spiegel-Rösing, Ina. "The Study of Science, Technology, and Society: Recent Trends and Future Challenges."

Storer, Norman. *The Social System of Science.* New York: Holt, Rinehart & Winston, 1966. Another discussion of the contemporary sociological view of the scientific community.

Szilard, Leo. *The Voice of the Dolphins and Other Stories.* New York: Simon & Schuster, 1961. A collection of essays, some in the form of parables, by a very original scientific mind.

Watson, James D. *The Double Helix*. New York: Atheneum Publishers, 1968. A personal account of the race to disentangle the structure of the DNA molecule.

Weinberg, Alvin. *Reflections on Big Science*. Cambridge, Mass.: MIT Press, 1967. Influential analyses by an American scientist with vast experience in practical science policy.

Ziman, John. *Public Knowledge*. Cambridge: Cambridge University Press, 1968. A noted British physicist emphasizes the importance of consensus in bringing about the objectivity of science.

Zuckerman, Harriet. *Scientific Elite: Nobel Laureates in the United States*. New York: Free Press, 1977. An extensive sociological study of prominent people in science.

Zuckerman, Sir Solly. *Beyond the Ivory Tower: The Frontiers of Public and Private Science*. New York: Taplinger, 1971. Comments by one of the great figures in British science policy.

INDEX